AF738361

Digital Twins

Sunil Gupta • Ravi S. Iyer • Sanjeev Kumar

Digital Twins

Advancements in Theory, Implementation, and Applications

Springer

Sunil Gupta
UPES University
Dehradun, Uttarakhand, India

Ravi S. Iyer
UPES University
Dehradun, Uttarakhand, India

Sanjeev Kumar
UPES University
Dehradun, Uttarakhand, India

ISBN 978-3-031-76563-6 ISBN 978-3-031-76564-3 (eBook)
https://doi.org/10.1007/978-3-031-76564-3

© The Editor(s) (if applicable) and The Author(s), under exclusive license to Springer Nature Switzerland AG 2025

This work is subject to copyright. All rights are solely and exclusively licensed by the Publisher, whether the whole or part of the material is concerned, specifically the rights of translation, reprinting, reuse of illustrations, recitation, broadcasting, reproduction on microfilms or in any other physical way, and transmission or information storage and retrieval, electronic adaptation, computer software, or by similar or dissimilar methodology now known or hereafter developed.
The use of general descriptive names, registered names, trademarks, service marks, etc. in this publication does not imply, even in the absence of a specific statement, that such names are exempt from the relevant protective laws and regulations and therefore free for general use.
The publisher, the authors and the editors are safe to assume that the advice and information in this book are believed to be true and accurate at the date of publication. Neither the publisher nor the authors or the editors give a warranty, expressed or implied, with respect to the material contained herein or for any errors or omissions that may have been made. The publisher remains neutral with regard to jurisdictional claims in published maps and institutional affiliations.

This Springer imprint is published by the registered company Springer Nature Switzerland AG
The registered company address is: Gewerbestrasse 11, 6330 Cham, Switzerland

If disposing of this product, please recycle the paper.

Foreword

Embracing the Digital Reflection: A Journey into the World of Digital Twins

Today, the world around us is changing dramatically—becoming intertwined with both the physical and the digital realities. At the forefront of this revolution lies a powerful technology: The digital twin is the term for this kind of real time, dynamic replica of a system. Digital twins, carefully explained in this book, act as a pipeline that helps organizations recognize, apply, and take advantage of the possibilities of digital twins. In these pages you embark on a voyage that explores the richness.

Chapter 1 provides the initial information and background necessary for comprehending the application of digital twin technology. By the end of the course, you will understand digital representation, virtualization, and the perfect balance of the physical and virtual world. However, it does not stop here at the theory.

Chapter 2 brings up certain ugly truths associated with the advent of digital twins. Potential risks concerning data security, privacy, quality, and technical issues are described; useful recommendations are provided.

Chapter 3 focuses on detailed description of activities that are undertaken. It reveals the concept of digital twins, the data process challenge, and the sequence of digital twin creation. Furthermore, it also looks at the emerging trends and practices that characterize the field. The beauty of digital twins is the variety.

Chapter 4 illustrates these through interesting industry case studies. Starting with manufacturing to the application of digital twins in healthcare, you are going to learn how digital twins are transforming different industries and the prospects and difficulties faced during the process.

Chapter 5 brings focus to one of the core dimensions of Digital Twins—Modeling. It reveals the elegant mathematical methods that are crucial for the practical creation of models or performances with high level of fidelity. More is yet to be expected in the future.

In Chap. 6, the author takes the reader through how artificial intelligence can enhance digital twin approach. Cyber Twin AI and its advantages for systems, which are its presence, bring in ideas like system optimization, anomaly detection, and failure prediction to the limelight.

Chapter 7 conversely gives the reader a bird's-eye view with actual examples across differing industries, from the ground level. Learn more about how companies are starting to use digital twins, with specific examples in aerospace, automobile, and the wellness sector.

Chapter 8 offers the final vision and looks ahead. This chapter reveals more on the further development of this revolutionary technology by providing information on future development, possible uses, and how it is likely to revolutionize the society and the environment.

This book is a go-to resource to get acquainted with the possibilities of using digital twins. Just submerge yourself and swim through the sea of possibilities where the physical and the digital are not two distinct worlds, but rather one where the two worlds are as one.

Former Professor & Head, Delhi College of Engineering Goldie Gabrani
Govt of Delhi Now Dean Engineering
and Executive Director Research Manav Rachna University,
Faridabad, Haryana, India

Preface

The idea of the Digital Twin—born just a few years ago—has changed the way that innovation, maintenance, and optimization are brought from conception to realization within industries. Born in the aerospace sector, sensitivity to precision and reliability, its growth has been rapid and is invading manufacturing, healthcare, transportation, and many other sectors around the world. This technology permits the creation of a virtual replica of a physical entity that reflects, in real time, its status, behavior, and evolution.

Digital Twin: Advancements in Theory, Implementation, and Applications intends to brief readers about the status of both theoretical background and the implementation strategies of this technology. During this evolution process, it is felt that the future challenges and opportunities for this technology will be led by an understanding of their current capabilities. These chapters have been meticulously selected and bring a wide bandwidth of topics to the fore—from basic concepts to state-of-the-art applications, and finally, the integration of Digital Twin Technology with current emerging trends such as Artificial Intelligence and Industry 4.0.

Chapter 1, "Digital Twin: Theory and Concepts," hence sets the foundation of this text by discussing the origin and critical theoretical aspects of the Digital Twin technology. It provides a clear insight into concepts such as digital representation, virtualization, and synchronization that are critical in the full comprehension of Digital Twins.

Chapter 2 of this book, "Modeling of Digital Twins," will introduce the reader to the most advanced modeling techniques forming the basis for the creation and operation of Digital Twins. This chapter presents the very different methods from physical and data-driven modeling to hybrid or multi-scale modeling, elaborating on how these models can enhance digital twins in terms of accuracy and usability.

Chapter 3, "Digital Twin Implementation: Architecture and Development," concerns practical implementation in terms of Digital Twin systems: architectural frameworks, data flows, and challenges related to developing such big and complex systems. Emerging trends and best practices going forward are also highlighted, providing insights into the future direction of implementation with Digital Twins.

Chapter 4 is about real-life case scenarios of "Industry-specific Digital Twin Applications and Use Cases" in which, within various fields, the technology has been pushed and implemented. The chapter aims to provide detailed case studies that demonstrate the real benefits and challenges that different industries face in their efforts to adopt and implement digital twins in their operations.

Chapter 5, Issues of Digital Twin, discusses all the challenges the technology is facing, including issues on security, privacy, and data quality. Besides, the technical challenges, for the successful deployment of digital twins in practice, are underlined.

Chapter 6, "Digital Twin and Artificial Intelligence," combines the technologies of Digital Twin and Artificial Intelligence. Here, the story will be set for how AI can bring functionality toward improving the performance of applications in Digital Twins to the forecasting of potential failure.

Chapter 7, "Digital Twin and Cybersecurity," covers the whole aspect of the security that inheres in the application of Digital Twins. It comprehensively describes the vulnerabilities inbuilt within Digital Twin systems to provide the best practices in safeguarding and ensuring privacy.

Chapter 8, "Digital Twin and Industry 4.0," correlates the Digital Twin technologies to the more general context of Industry 4.0. In situ realization, real-time monitoring, and prediction of maintenance with Digital Twins are major features of the fourth industrial revolution.

The book identifies real-world deployment cases of Digital Twins; thus, there are lessons and insights of immense value from an industrial point of view in areas such as aerospace, automotive, and healthcare. The presented case studies talk about challenges and successes while implementing Digital Twins. Lastly, Chap. 8 "Future Directions of Digital Twin Technology," also provides a prediction of what can be expected in terms of innovations and applications within the field of digital twin technologies. It presents discourses pertinent to emergent trends and their impact on society, coupled with the role that digital twins will play in the future.

Dehradun, Uttarakhand, India

Sunil Gupta
Ravi S. Iyer
Sanjeev Kumar

Contents

Chapter 1
Digital Twin: Theory and Concepts

1.1 Introduction

The digital twin technology essential concepts are discovered by the theoretical analysis. The chapter examines the nuances of digital representation and in addition it shows how meticulously real-world objects are replicated online. Subsequent, virtualization is the process of making virtual copies of real-world object and it is used to examine the ability to bl the virtual and the real. This book discovers and elaborate the development of data exchange and real-time communication between physical and virtual systems. Which is essential to the digital twin's operation. In this chapter, the digital twin technology, and a deep understanding of its principles are enlightening that open up the surfeit of potential within the digital ecosystem [1].

1.1.1 Definition of a Digital Twin

The Fig. 1.1 shows a digital twin representation of a real-world object, person, and procedure in a digital duplicate and its surroundings. Digital twins help us in making better decisions and can be used by an organisation to pretend actual experiences and their results. This behaves as dynamic replica, a data-driven, that simulates the characteristics, behaviours, and standing of its real-world counterpart To provide accurate real-time reflection of the physical object, the method uses a variety of data sources, including computer models, sensors, and Internet of Things (IoT). The digital twin can contain the infrastructure of an entire city, as well as the individual machinery and equipment, as a whole industrial facilities.

The bidirectional communication and synchronisation between the virtual and physical realms is the primary features of a digital twin. For the reason that, changes or events that take place in the real world are mirrored in the digital equivalent and vice versa. Using the technology the operators, engineers, and decision-makers can

© The Author(s), under exclusive license to Springer Nature Switzerland AG 2025

S. Gupta et al., *Digital Twins*, https://doi.org/10.1007/978-3-031-76564-3_1

Fig. 1.1 Digital Twin

gain deeper insights, make informed decisions, and even predict future behaviour. The technology makes it possible to analyse, monitor, and operate physical systems in real time. As digital twin technology advances and becomes more mature, it is fundamentally changing the way we design, supervise, and optimise the complex systems. Healthcare, manufacturing, transportation, and smart cities are a few instances of these industries [1].

Digital twins are updated continuously to match the original since they are linked to real environmental data sources. A digital twin is an online replica of an actual entity that may be used to mimic its actions and get additional insights into how it operates. To improve simulations, scenario planning, and decision making, an enterprise metaverse is a digital, usually immersive environment that mimics and integrates every part of a business. Digital twins also contain a layer of behavioural insights and data-driven visualisations. When digital twins are networked within a single system, it becomes feasible.

There are various types of digital twins. The first kind is called a product twin, which is a illustration of a product. It include products at different stages of their life cycles, from early concept design and engineering to complete operation, providing you with live, and real-time data on a product as if it were in use. An google map is one of the examples of a product twin that you probably already have in your pocket all the time: They connects you to real-time traffic information to make travel more efficient.

1.1.2 Benefits of Digital Twins

In the digital age, digital twin technology is essential because it provides a multitude of benefits and opportunities that profoundly affect numerous facets of our lives and businesses. The following are some main justifications for its importance:

- Enhanced Performance and Efficiency: Real-time monitoring and analysis of physical systems is made feasible by digital twins, which enhances performance and efficiency. With the aid of an accurate digital depiction of a physical object, operators and engineers may identify issues, make data-driven adjustments, and optimise processes, improving productivity and reducing operating costs.
- Predictive Maintenance: By digital twin, they give an enterprise continuous observation into the condition and activity of its physical assets. One can identify with ease the potential problems before they are fateful enough to cause expensive breakdowns by analyzing sensor information and performance history of the asset, saving downtime, and avoiding unwarranted failures.
- Informed Decision-Making: Digital twins enable decision-makers to access the latest intelligent data. With the right data about the physical system's condition, any organization working in manufacturing, health, or even urban planning will make the right decisions, resource things properly, and flexibly respond to situations.
- Better Innovation: Simply, the creation of a "digital twin" vis-à-vis new products and processes is a critical philosophy. Offering organizations the ability to try out ideas in the virtual world before making changes in the real world will decrease the risks associated with innovation and bring down development cycles.
- Personalization: Individual treatment plans can be developed in areas such as healthcare based on the digital image created with respect to the unique physiology of a patient. Such a personalized action goes on to enhance the care and outcomes of the patients.
- Environmental Impact: Optimizing the energy use and environmental impact of physical systems can be done through digital twin technology. This will turn out to be very critical in reducing the carbon emission and encouraging sustainable urban development.
- Remote Operation and Collaboration: A digital twin enables experts to collaborate on systems from any location around the world. Recent experiences have shown this to be very useful as it allows continuation of crucial operations in situations like the COVID-19 pandemic.
- Improved Safety: Digital twins, through simulation and real-time monitoring, make many systems such as highly complex industrial processes and driverless cars—much safer. It is quite possible to identify and thereby avert probable risks in an effective manner.
- Big Data and Analytics: One of the most important aspects of the creation of digital twins is related to the large volumes of data that come as part of it. This has created an independent sector of data analytics growth. All this data can be harnessed for newer insights, trend analysis, and even used for developing machine learning algorithms for further enhancing the performance of the system (Fig. 1.2).

1.1.3 How Do Digital Twins Contribute to the Environmental Sustainability of an Organisation?

The digital twin products greatly assist in performing sustainable development. These digital twins would enable firms to decrease the amount of material usage during the design phase of products and enhance the traceability of products to minimize waste generated in the environment. After implementing digital twins, consumer electronics manufacturers have achieved a steep growth in sustainability.

Digital twins can be used to monitor in real time, give predictions through analytics, and provide data insight for optimizing resources, increasing energy efficiency, and reducing emissions all factors that can significantly help toward environmental sustainability. They do this by identifying improvement opportunities, reducing resource usage, cutting down on waste, and accordingly enhancing the overall efficiency of energy consumption through virtual duplicates of physical systems and processes. These results in reduced emissions, better efficiency within supply chains, and other practices that reduce greenhouse gases. Digital twins also help in reducing electronic waste, lengthening the lifecycle of assets, and attaining compliance through a reduction in unwarranted maintenance. All these are supposed to earning business commendations for digital twins to make informed eco-secure decisions that foster commitment towards sustainability, consequently aligning corporate goals with sensitive environmental concern. Digital twins can optimize improvement in environmental sustainability through improved resource efficiency and reduction of wastage, and contribute to more environmentally friendly practices. DT can influence the environmental sustainability of any organization in the following ways: [2, 3].

- Energy Efficiency: Digital twins can aid in optimization and modeling of organizational energy consumption. Provision of the virtual models of infrastructures, factories, and buildings enables real-time monitoring and analysis of its energy use. Further, adjustments can be made on Heating, Ventilation, and Air

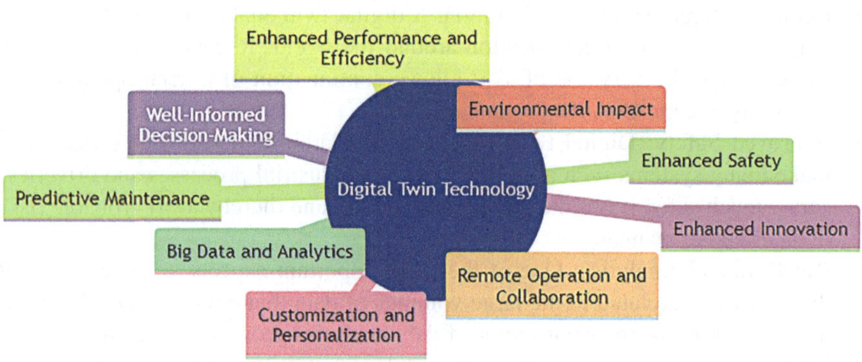

Fig. 1.2 Benefits of Digital Twins

Conditioning Systems, lighting, and other elements consuming energy to avoid this waste and reduce energy expenses. This also reduces the carbon footprint apart from saving money.

- Resource Efficiency: Digital Twins can provide insights on how to better make use of resources in terms of water, raw materials, and source materials. This could manifest in the manufacturing industry, for instance, in a digital twin that models the production process and makes suggestions on how to lessen waste material and reduce usage of resources by enabling the use of more sustainable sources. Others include agricultural digital twins, which would assist in optimization of irrigation systems, thus making sure that water is utilized effectively. Organizations contribute to sustainable development by ensuring minimal wastage of the available resources.
- Supply Chain Efficiency: Organizations can use the digital twin of the supply chain to minimize inventory wastes and make necessary changes to reduce emissions through transportation. It provides predictive analytics, load balancing, and optimized routing that reduces fuel consumption and, thereby, can reduce GHF emissions. Decisions about sustainable sourcing and transport choices can be made more accurately.
- Emissions Reduction: Digital twins of the organization's emissions tell where improvements are necessary. This would be by modeling and optimizing their manufacturing, reducing fuel usage within transport, and more rigorous management of carbon offsetting. All these can be realized through process simulation and activities, hence reducing the environmental impact of the organization and enabling it to meet its emission reduction goals.
- Smart Buildings: The digital twin of buildings makes it possible to enable smart management processes of buildings for real estate and property management companies. This includes real-time observation of lighting, HVAC systems, and energy use. By changing factors such as occupation and others, these allow for more environment-friendly building management practices and the enabling of energy saving.
- Predictive Maintenance: By predicting which equipment is likely to fail and undertaking maintenance tasks accordingly, organizations can avoid all such waste of unwanted repairs/replacements. It extends the life of the assets, besides reducing the quantum of mechanical and electrical equipment that has to be sent to landfills, improving sustainability.
- Environmental Compliance: Digital twins can help organizations ensure that environmental laws and regulations are adhered to. With reporting and addressing of environmental concerns related to emissions, waste management, and resource usage in good timing, organizations can avoid hefty fines and show a serious commitment to the practices of sustainability.

1.1.4 *The Digital Twin's History*

The idea of producing virtual representations of actual elements and systems has its roots in several domains, which is where the concept of the "digital twin" first emerged. The comprehensive approach we know today was founded on these early discoveries.

Michael Grieves first discussed digital twins in an early 2000s course presentation on product lifecycle management. 2011 saw the implementation of digital twins as a challenging process requiring numerous technological advancements [4]. Although the term "digital twin" was first used in 2003, NASA's Technology Roadmaps published an explanation of the concept years later. In these descriptions, a hardware twin was utilised to replicate space conditions and conduct tests in advance of takeoff (another example of such hardware twin was the Airbus Iron Bird). The DT technique began with the aerospace sector and was later extended to the manufacturing sector in 2012. Numerous fields, including Industry 4.0 [5], the Physical Internet, Cyber-manufacturing, and Cloud Manufacturing, have witnessed significant developments as a result of the breakthroughs in technologies like cloud computing, IoT, and big data. Industry 4.0 has undergone a change mostly due to big data, IoT, and digital improvements. The concept of digital twins was revitalised due to Industry 4.0, which allowed for the digital storage of all data and the integration of sensors into industrial areas. It also became feasible to test realistically in a virtual environment because the high increase in processing resources availed advanced simulation capabilities. Functional digital twin as a utility for companies like IBM, Siemens, and even GE started getting deployed for their own use and for their clients as a function of these technological advancements [6].

NASA and Aerospace
The concept of digital twins is hugely contributed by the aircraft sector. NASA began simulating and testing spacecraft and their constituents on computer models and simulations in the middle of the twentieth century. In space exploration, where an unfriendly environment is required in space, these early digital twinning techniques were of essence as they allowed engineers to understand how the systems would work. This way of designing created safer missions and more efficient designs.

Manufacturing with Sector 4.0
The manufacturing sector has been a strong sector that really influenced the development of digital twins. From the definition given for Industry 4.0, which introduces the incorporation of digital technologies in manufacturing, more advanced digital twin concepts were developed. The manufacturers initiated the modelling of their production lines, equipment, and goods virtually. The digital twins allowed for real-time monitoring with predictive maintenance, thus improving productivity and reducing downtime [7].

IoT and Smart Cities Digital twin technology was once again on the rise with the coming of the Internet of Things and increasing smart cities. The Internet of Things connected actual things and infrastructure to the internet, generating enormous data volumes. It is now possible to simulate and optimize most facets of city life, from public safety and healthcare to transport and energy management, by using full urban environment digital twins.

These early developments in digital twin technology in manufacturing, aircraft, and smart cities helped set the base for its wide application across different industries. The idea has since evolved into a unifying theme that spans these diverse fields and allows for the integration of real-time data and virtualization of complex systems to enable a better understanding of them.

1.1.4.1 Technological Development of Digital Twins Over the Years

The history of digital twins, from their very basic dawn to their full-scale application across versatile industries in modern days, can be traced back to some important milestones. From early uses in manufacturing and aerospace to the pervasiveness of technology in this age of the Internet of Things and Industry 4.0, digital twin technology has taken its course in history, as indicated in Fig. 1.3. It also influences how we interface with, and optimize, physical systems in the digital age.

1960s–1980s: Early Simulations and Aerospace
These are the fields where the concept of digital twins first emerged. Initially, NASA and the aerospace industry ran computational models of spacecraft and aeroplanes, simulating them and then running several tests. Early examples of digital twinning like this helped engineers (albeit only partially) to learn how the physical systems may respond under different circumstances, enabling them to enhance the design and ensure safety in aeronautics and space exploration.

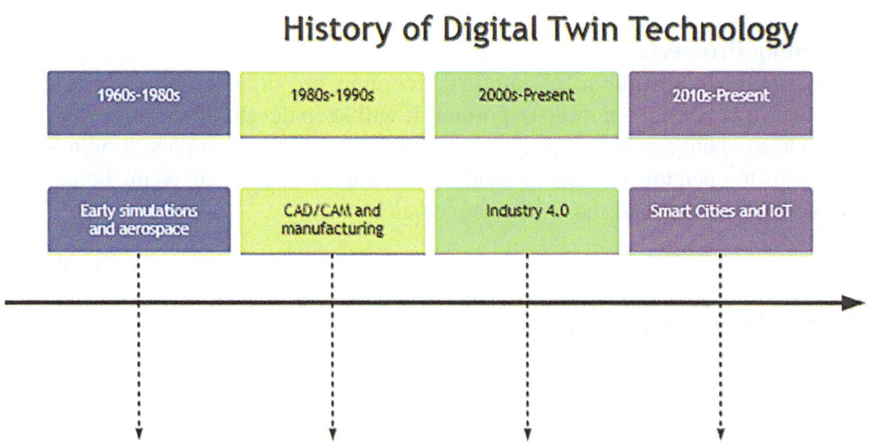

Fig. 1.3 History of digital twin technology

CAD/CAM and manufacturing (1980s–1990s)
It was during this period—in about the 1980s and 1990s—that the central turn towards the development of digital twins, with the introduction of CAD and CAM, took place. These systems provided for constructing digital replicas for physical products and manufacturing processes. That meant laying the basis for virtualization and digital representation of manufacturing systems to increase production efficiency.

Industry 4.0 (2000s–Present)
The development of digital twin technology has been accelerated by the emergence of Industry 4.0, which is defined by the incorporation of digital technologies into industrial and manufacturing processes. It presented the idea of the "smart factory," in which digital and physical systems are nearly identical. Production process optimisation, predictive maintenance, and real-time monitoring are made possible by these digital twins. As a result, productivity has grown, downtime has decreased, and product quality has improved.

Smart Cities and IoT (2010s–Present)
As the Internet of Things (IoT) grew, digital twins' applications went beyond manufacturing. The internet of things (IoT) enabled the connection of infrastructure and physical objects to the internet, producing enormous volumes of data. Utilising this data, digital twins of complete systems—like smart cities—were produced. Digital twins are currently used by urban planners to model and improve many facets of city life, including public safety, healthcare, energy management, and transportation.

Current Cross-Industry Adoption Digital twin technology has become increasingly popular in a number of areas, including healthcare, energy, transportation, and agriculture. Precision agriculture, energy grid optimisation, driverless vehicle testing, and healthcare simulations are just a few of the uses for it. The possibilities of digital twins are further enhanced by the ongoing development of sensors, data analytics, and simulation tools.

Upcoming Prospects
With the digital revolution of industries coming up, digital twin technology is expected to become even more important. It will keep developing and, in so doing, continue to change how we design, manage, and optimize complex systems—the new paths to open toward environmental sustainability, personalized medicine, predictive analytics, and all those other huge fields.

1.1.5 Theoretical Foundation

A number of essential ideas, which get borrowed from other domains and form the theoretical base of the technology, prop up the development and operation of digital twin technology. Digital twin technology fundamentally falls back on the basic concepts of cyber-physical systems. CPS studies the interaction of computational and physical components with a focus on real-time control and communication. Digital twins act as a connecting link between the real world and its digital counterpart to maintain this bridge and thus embed smooth synchronisation. Due to this theoretical base, interaction with physical entities and copying them is also possible with digital twins. This again allows receiving real-time insights and enables the possibility of remote control and optimisation.

Modelling and simulation theories are also embedded in digital twin technology. This calls for sophisticated modeling techniques that will be able to give an adequate representation of the behavior, structure, and features of the modeled physical object. In this way, this digital twin modeling enables tests of various scenarios and "what-if" evaluations, hence supporting system design and decision-making. Machine learning and data analytics could provide more advanced abilities in digital twins. Digital twins are of importance to any business relating to precision, efficiency, and data-driven decision-making; they might give predictive insight and adaptive control through constant collection and analysis of data from both physical and virtual reality. Theoretically, the technology of digital twins can be viewed as the result of convergence between modeling and simulation, data analytics, and cyber-physical systems. The convergence of these produces dynamic, real-time digital twins that foster innovation and optimisation in multiple dimensions.

1.1.5.1 Digital Twin Development

Digital twins are underpinned by concepts from several domains: data analytics, simulation, and cyberphysical systems. The contribution of each domain in relating to is as follows [8]:

Cyber-Physical Systems: The ideas of CPS consider seamless integration between computational and physical elements, hence creating the foundation for a digital twin. CPS places an emphasis on real-time control and communication between digital and physical elements. That is, digital twins maintain two-way connections with their physical twins; changes made in the physical system are seen in the digital twin and vice versa. In this manner, digital twins become very useful tools in monitoring, control, and decision support in industries such as manufacturing and healthcare to smart cities, because of their real-time synchronisation that allows dynamic interaction.

Simulation: This digital twin technology, however, majorly relies on techniques of simulation for building and managing the virtual representations of some non-virtual objects. Simulations make use of computational models that try to emulate characteristics, responses, and behaviors of real-world systems. It is imperative for the correct, faithful representation of physical entities in the digital sphere. Using simulations, digital twins can predict the behaviors of real-life systems when varying conditions operate within them, hence giving the opportunity for a decision to be well-advised ahead of time. They help reduce the risks that often accompany novelty and change in the real world by offering experimentation in the virtual environment. Consequently, simulation assumes a very significant role in the design, testing, and optimization of a digital twin.

Data Analytics: Digital twin is much dependent on data analytics. Digital twins continuously monitor and collect data concerning the virtual and physical worlds. It contains, but is not limited to, sensor readings, ambient conditions, and other relevant data. All this information is finally fed into data analytics techniques to infer, recognize patterns, and predict outcomes. This has become more crucial to enterprises seeking to obtain maximum productivity and efficiency, predict maintenance needs, and make data-driven decisions. Running machine learning algorithms against extensive datasets gives patterns that would prove laborious to determine through research manually. In simple terms, data analytics empowers the digital twin as a dynamic, evolving system that offers hands-on insight to its users.

1.1.5.2 Digital Replica of the System or Physical Object

It creates a true virtual replica of the physical object or system to be monitored, analyzed, and interacted with in real-time. The procedure will thus be part of the essential core of digital twin technology, helping as a driver to bridge the gap between the digital and the physical world. The mechanism of operation for the Digital Representation is as follows: [9].

- Acquisition of Data: This is the collection of data from an existing object or system. To a large extent, it can be collected through human entries, IoT devices, cameras, and sensors. The sensors capture the position, orientation, temperature, pressure, and other relevant properties regarding the object. This information gathered forms the basis upon which the digital twin is created.
- Graphics and Modeling: A digital model of the real object or system under consideration is created with the grabbed information. It may need a mathematical model, a 3D representation, or a combination. The model captures geometry, characteristics, and behavior of the real thing. Later, this digital model will be rendered using visualization techniques to create a virtual representation nearly identical to the real version.
- Real-Time Synchronization: This is an important part related to the working of digital twins. This means that any changes in the physical system will be reflected in the digital twin and vice-versa. For example, every change in the physical object, like movement or a change in temperature, is changed in the digital copy in real-time. On the other hand, if it is a physical system built for bidirectional control, then any changes in its digital twin might affect it.
- Data Integration: The digital twin requires data integration so that its digital representation is valid and up-to-date. Data from various sensors and other sources feeds into the digital twin as a representation of what is really happening. It changes in real-time as the physical system changes. Data integration is often

supplemented by strong, in-built data analytics and real-time processing, which aims to gain meaningful insights from information gathered.

- Interactivity: After the digital twin is in place, a user can interact with it to monitor or analyze the physical system. Since it offers interactivity, users are allowed to visualize an object's real-time status, run several kinds of simulations, execute "what-if" assessments, or make defendable judgments with real-time data.

Digital twin technology is hardwired into a digital representation; it's a process meant to replicate faithfully the physical systems in the virtual world. Due to their fidelity and real-time synchronisation across the two realms, the digital twins become critical tools for monitoring, analysis, and optimization of complex physical systems across sectors.

1.1.5.3 Method by Which Virtual and Physical Systems Exchange Data

Since virtualization is the base concept of digital twin technology, it mainly deals with making a virtual copy or equivalent of devices/systems/objects/units, or even things or entities, involving real-world systems. This approach will enable the real-time digital simulation and mirroring of the physical environment. The process of creation of the virtual equivalent for a real-world object is shown below and in Fig. 1.4 [9]:

- Data Acquisition and Capture: Virtualization of an object or system starts with the collection or gathering of data about the physical object or system. This can be sourced from many sources, such as sensors, cameras, GPS, environmental monitors, and any other device down the extensive list of all data acquisition tools. These sensors monitor data concerning the position, velocity, temperature, pressure, or any other characteristic of the physical thing and its conduct and surroundings.

- Data Processing and Transformation: The Campaign datasets that were collected are then processed and transformed into a digital format, which will enable the creation of virtual models. During the transformation process, the raw sensor data is converted into a digital representation of the shape, characteristics, and behaviour of the physical entity. The dynamics of a physical system, very often expressed in a mathematical form, support the delivery of correct simulations. These models should very closely mimic real-world system behaviour.

  - Simulation and Visualisation: After establishing the virtual model, the following stage is generating a virtual counterpart through simulation and visualisation methods. A digital environment, such as computer-aided design (CAD) software, 3D visualisation, or other It utilizes specialized tools to render the

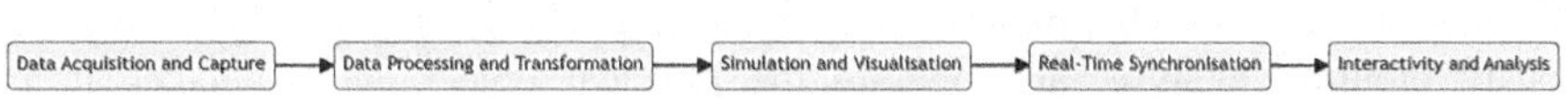

Fig. 1.4 Producing virtual equivalents of real-world objects

model. This virtual representation makes it possible for a real-time visual depiction of an object or any system by emulating the appearance and qualities of an overrun entity.

- Real-time Synchronization: The virtualization mentioned above will allow real-time synchronization between the physical object and its version existing in the virtual world. It means that any activity or change taking place in reality will automatically be pushed to the digital twin, and vice versa. In this way, to depict the very aspect of preciseness and views updated at runtime, the virtual replica will modify according to the changes taking place on the actual thing, like movement, heating up, transformation etc.
- Interactivity and Analysis: The virtual equivalent can now be interactively worked on and analyzed by users, engineers, and decision-makers. The simulations can be run, and forecast studies are carried out; the current status of the real system monitored, with decisions taken by using data from it. One is at liberty to conduct experiments and "what-if" situations, if desired, without affecting the real physical thing through virtualization.

Virtualization enables digital twins, which means it offers a dynamic and real-time link between the digital and physical worlds. This ability is to optimize complex systems in the physical world and run industries, from manufacturing and healthcare to transportation and smart cities. At the same time, this is offering insightful information that improves decision making.

1.1.5.4 How Virtual and Physical Systems Will Be Connected, and How Data Will Be Exchanged Between Them

Synchronization is an essential process in digital twin technology since it will provide for interchangeability of data and real-time connectivity between virtual and physical systems. It licenses dynamic crossbreeding whereby actions or a state change in reality is instantaneously represented within the digital double and vice versa. It makes it a real-time connection through constant data interchange and communication, improving the digital twin's capability of producing an exemplary and current representation of the physical entity.

In the synchronisation process, data is exchanged in both directions between the physical and virtual systems. Real-time data acquired from sensors and Internet of Things devices in the real world is communicated to the digital twin. This information contains the condition, state of being, and other features associated with the physical object. A digital twin processes it and contributes to its data from the virtual model. This data could be produced by sensor readings, changes in the environment, or a change in operations. The digital twin transmits all the changes made with it during that time back to the real system once again, such as control orders or simulated situations.This bidirectional connectivity makes it possible to continuously monitor, analyse, and operate the physical object, which makes it an invaluable tool for predictive maintenance, real-time decision-making, and performance optimisation in a variety of sectors [10].

1.1.6 Use Cases and Applications

The technique has a wide range of applications because it can create digital twins of particular parts, entire assets, entire systems, and entire processes.

New systems are tested before being manufactured.

Before investing in construction or implementation, businesses can utilise digital twins to design and evaluate systems, equipment concepts, and service models. In the event that a model is successful, real-time monitoring may possibly be facilitated by connecting its digital duplicate to the actual creation.

Increasing Production and Efficiency
According to a 2017 Forbes estimate, the use of digital twins might result in a 30% increase in the pace of crucial processes. Gartner estimates that industrial enterprises may increase their effectiveness by 10%. Digital twins are widely available and have a wide range of applications, which helps companies in almost every industry identify areas where processes may be optimised and streamlined. This is achieved through predictive maintenance that lowering cases of downtime.

Real-Time Asset Management
By optimizing production and monitoring of daily activities, Digital Twins prevent unwarranted stress put on the equipment and notify involved company owners of cost-friendly modifications, such as adjusting fuel consumption. Organizations remain competitive with the general increased production whereby maintenance and repair operations are done quickly.

Knowing Data to Deliver Better Services
Besides this, digital twins can also be utilized in a customer-facing environment, such as remote troubleshooting. Instead of following blindly through the default protocols, personnel could conduct diagnostic tests using virtual models from any location and lead customers through the right procedures to make the necessary repairs. Data collected from these sessions gives useful insight into planning and designing future products.

The technology of digital twin has been in use in many industries, completely changing how companies operating in those sectors used to operate and make decisions. It finds application in a few other industries besides the following: healthcare, automotive, and smart cities [11]:

1. = Medical Care:

   Personalized Treatment Plans: Digital twins, replicating each patient's physiology, are used to create virtual copies of patients. Physicians are able to utilize these personalised digital twins to develop optimized treatment programs for conditions such as cancer or cardiovascular diseases, or design a treatment program based on the needs and pharmaceutical interactions of each patient.

2. Surgical Simulations: Before real surgeries, surgeons practice on challenging procedures in digital twins. Ultimately, augmented risk control and increased surgical accuracy lead to better patient outcomes.
3. Vehicles:

   The need for digital twins in autonomous vehicles arises for testing and improving autonomous systems. This will assure the performance and safety of self-driving vehicles through modeling various driving situations in protected virtual test environments, without risks in the real world.
4. Predictive Maintenance: The automakers are using digital twins to trace and estimate the maintenance requirements of their cars. These virtual representations bring down the frequency of car breakdowns and its related expenses by helping in detecting problems at an early stage.
5. Smart City:

   Urban Planning: Whole city digital twins are used to maximize urban planning. Some such representations include public services, energy networks, or transportation systems. With those, urban planners will be able to advance sustainability, make well-informed decisions around infrastructure, and simulate the impact of planned changes.
6. Citizen Safety: Digital twins applied in smart cities realize the life-critical monitoring arrangements over public places and infrastructures. A very nice example here is the faster detection of, and timely response to, environmental hazards, security breaches, or accidents.
7. Fabrication:

   Factory Optimization: Manufacturing plants conceptualize their production lines and equipment in digital twin forms to enable predictive maintenance and real-time monitoring, thereby achieving reduced downtime, higher efficiency, and cost reduction.

   Product Development: Industries in the automotive, aerospace, and consumer electronics sectors use Digital Twins to simulate and test new product designs. This reduces the creation of prototypes in reality and shortens development cycles [12].
8. Utility and Energy:

   Energy Grid Management: Utilities can monitor the distribution of electricity and optimize it through digital twins of their power systems. This reduces power disruptions, manages supply and demand, and integrates renewable energy sources into the grid with a high success rate.
9. Water and Waste Management: Toward this resource efficiency, municipalities utilize digital twins to perform waste collection, water supply network optimization, and water treatment facility management.
10. Farming

    Precision Farming: Digital twins of their fields and farms will help farmers maximize crop management through satellite and sensor data-based decisions on planting, irrigation, and harvesting.

Digital twin technology acts as one of the transformational tools across multiple disciplines due to its adaptability. Broad acceptance throughout these various industries and beyond is attributed to its real-time insights and predictive analyses and the capability for informed decision-making in general.

1.1.7 Implementation Issues of Digital Twin Technology

Whenever a new technology comes into limelight, security is always the last thing considered in tr he initial stages. Weeks and months are lost in an attempt to extinguish figurative flames of weaknesses. On relative terms, compared to other types of systems, security risks for digital twins are low since they are cloud-based and do not Summand physical infrastructure. But the huge volumes of data that are being acquired and consumed come from multiple destinations, any of which may be a source of vulnerability.

While this use of digital twin technology holds tremendous potential, there are also pitfalls. Some of those present obstacles to utilizing this very revolutionary technology [11] include:

Data Integration and Security: The major challenge in the prominent position is handling volumes of data that a digital twin generates. This process involves integration from prime sources like databases, sensors, and Internet of Things devices, which may result in a resource overload. The integrity of a digital twin is under severe threat from cyber threats; hence, securing this data in a way to ensure its security and privacy is equally important. Protection from undesirable access and leaking of data is never a single step process, especially in information-sensitive industries such as healthcare and defence.

Interoperability and Standardisation: Interoperability and standardised work processes may not be achieved because different companies and applications require customized solutions for their digital twins. It could get problematic when digital twins are supposed to function smoothly with cohesiveness integrated into dissimilar software, hardware, and systems. Although much effort has gone into standardization, the variety in technologies and use cases involved makes it rather a challenging process. Realising full potential of the digital twin would require assuring that it is collaboratively exchanging information with efficiency and timeliness.

Expense and Investment in Resources: The implementation of digital twins may require huge financial and human-resource investments in technology structure and knowledge. The creation and sustaining of an exact digital depiction can be extremely costly where complex systems are involved. For smaller organizations, it may be difficult to get started with the associated costs and probably will not be in a position to use the technology to its fullest potential.

Change management: Implementing digital twin technology often involves changing the entire organizational culture. Workers need to adapt to new processes and learn to accept data-driven decision-making. Overcoming this resistance to change and arranging for the proper training of staff could become quite a challenge.

Basically, one should be aware that the quality and accuracy of a digital twin are member-based on the quality and relevance of the data used. In other words, it is always a battle concerning relevance, quality, and correctness of data used in the case of a digital twin. There could be some calibration issues, data bias, or even sensor errors that would impact how reliable the virtual representation is going to be.

Regulatory and Ethical Issues: Industry players need to deal with complex regulatory environments, from healthcare to self-driving cars. On a more specialized level in the application of digital twins, challenges pertaining to compliance with existent laws and the resolution of new ethical issues regarding data ownership and permission have been raised.

Scalability: As technologies based on digital twins become more and more diffused, scalability of systems and processing of constantly increasing data sets are gaining interest.. For digital twins to be successful, they must be able to manage the volume and complexity of data that is growing.

1.2 Conclusion

This chapter covered the history of digital twins include manufacturing and aerospace that shows an n in-depth examination of digital twin technology, illustrating its theoretical foundations, historical development, the introduction of Industry 4.0, and their development and the core principles we have explored the abstract foundations that support its operation and clarified the fundamental ideas of virtualization, synchronisation, and digital representation. The chapter includes the uses of digital twin technology in a variety of industries that includes the autonomous cars of the automotive industry, the intelligent urban spaces of smart cities, and personalised treatment plans in the healthcare industry. Furthermore, the adoption of digital twins has been accompanied by some hurdles, such as significant resource commitment, the need for a culture shift, and worries about data security and calibration. The competence of digital twin technology is to provide data-driven decision-making, predictive analytics, and real-time monitoring.

References

1. Sharma A, Kosasih E, Zhang J, Brintrup A, Calinescu A, Twins D (2022) State of the art theory and practice, challenges, and open research questions. J Ind Inf Integr 30:100383
2. Bolender T, Bürvenich G, Dalibor M, Rumpe B, Wortmann A (2021) Self-adaptive manufacturing with digital twins. In: 2021 international symposium on software engineering for adaptive and self-managing systems (SEAMS), Madrid, Spain, 2021, p 156–166. https://doi.org/10.1109/SEAMS51251.2021.00029

3. Carvalho R, Rodrigues A, da Silva. (2021) Sustainability requirements of digital twin-based systems: a meta systematic literature review. Appl Sci 11(12):5519. https://doi.org/10.3390/app11125519
4. Tuegel EJ, Ingraffea AR, Eason TG, Spottswood SM (2011) Reengineering aircraft structural life prediction using a digital twin. Int J Aerosp Eng 2011(1):1–14
5. Yang L (2017) Industry 4.0: a survey on technologies, applications and open research issues. J Ind Inf Integr 6:1–10
6. Schweiger L, Barth L (2023) Properties and characteristics of digital Twins: review of industrial definitions. SN Comput Sci 4:436. https://doi.org/10.1007/s42979-023-01937-4
7. Sjarov M et al (2020) The digital twin concept in industry – a review and systematization. In: 2020 25th IEEE international conference on emerging technologies and factory automation (ETFA), Vienna, Austria, 2020, p 1789–1796, https://doi.org/10.1109/ETFA46521.2020.9212089
8. Tao F, Qi Q, Wang L, Nee AYC (2019) Digital twins and cyber–physical systems toward smart manufacturing and industry 4.0: correlation and comparison. Engineering 5(4):653–661
9. Jeong D-Y et al (2022) Digital twin: technology evolution stages and implementation layers with technology elements. IEEE Access 10:52609–52620. https://doi.org/10.1109/ACCESS.2022.3174220
10. Geyda A, Fedorchenko L (2022) Digital twins of activities: role of information actions. In: 2022 32nd conference of open innovations association (FRUCT). IEEE, p 102–111. https://doi.org/10.23919/FRUCT56874.2022.9953805
11. Zhang Y, Fang L, Deng H, Qi Z, Liang H (2023) Recent advances and future perspectives of digital twins. In: 2023 IEEE international conference on control, electronics and computer technology (ICCECT), Jilin, China, 2023, p 1563–1566, https://doi.org/10.1109/ICCECT57938.2023.10140652
12. Lowenstein D, Mueth C (2022) Implementing a digital twin, design and test, test and measurement strategy. In: 2022 IEEE Autotestcon, National Harbor, MD, USA, 2022, p 1–6. https://doi.org/10.1109/AUTOTESTCON47462.2022.9984739

Chapter 2
Digital Twin Challenges

2.1 Introduction

Digital twins signify the virtual phase of an actual system or processes portraying real-time intelligence, forecasting power, and optimization potential. However, with this cutting-edge technology increasingly used by businesses, they face several challenges at times that hamper them from reaching full potential. Particularly, this chapter elaborates on a few challenges the organizations face during the adoption of digital twins and describes this transformative process [1].

Of these, the first is privacy and data security. A digital twin relies on a constant flow of data, always coming in from various sources, most of which contain private or sensitive information. The overwhelmed protection of the data is quite difficult, but doing so is very essential. Security lapses and privacy violations can result in serious consequences, from data breaches to legal repercussions. Apart from the key concerns, it outlines suggestions on how to improve data security and make sure compliance with privacy laws is in order, so that businesses may work liberally with digital twins, with data security kept in mind.

Secondly, data quality is another major problem described in this chapter. For digital twins to deliver on their promise, the data required to create and keep these virtual minions must be reliable, repeatable, and accurate. On the other hand, data quality might turn bad due to some factors like data anomalies, sensor errors, etc. Poor data quality curbs the ability to make decisions renders forecasts less powerful, and ultimately impacts the overall quality of experience. To tackle this challenge, we investigate techniques and processes for enhancing data quality to ensure it as the linchpin of the digital twins is proven and improved.

The third dimension of the challenges is composed of technical problems such as integration complications, scalability concerns, etc. Data comes in a wide variety of shapes and sizes, from different systems, devices, and sources and this can be challenging to bring together, and as digital twin applications become more and more common, scalability is crucial. This chapter discusses the technical nuances

© The Author(s), under exclusive license to Springer Nature Switzerland AG 2025

S. Gupta et al., *Digital Twins*, https://doi.org/10.1007/978-3-031-76564-3_2

surrounding the implementation of digital twins to help organizations maximize the benefits of digital twins in various industries, addresses best practices, and introduces potential next-generation technologies that might help scale these technology challenges.

In this chapter, we focus on giving readers a holistic vision of how deep-rooted and multilayered the challenges around the adoption and implementation of digital twins can be by touching on these 3 dimensions of challenge—data security and privacy, data quality, and technical. It also provides practical advice for organizations to navigate and overcome these challenges, so the digital twins promise can be met in full.

2.2 Digital Twin Adoption and Importance Across Industries

Digital twins are one of the latest crazes that have been adopted in various industries over the last few years, revolutionizing the way businesses interact with and perceive their physical assets. Digital Twins of even become an integral part of modern industrial and commercial landscapes scaling the bounds of aerospace and engineering to industrial, manufacturing, healthcare, MQTTs, and urban planning being recent applications in the IoT world.

This is the crux of the matter because this leads to the unique ability to bridge the digital and the real, thus, creating a composited live anima of the real objects, which are, to the best of my knowledge, accurately similar to the real models or prototypes. This symbiosis has opened up a new innovative dimension in efficiency and effectiveness, and businesses are keen to leverage the potential of applying the concept of digital twins to differentiate value chains and enhance the decision-making framework.

The attraction of digital twins is their capacity to produce a data-driven, real-time replica of physical assets, providing an unprecedented level of detail. Digital twins, for example, are transforming patient care in the healthcare industry by enabling physicians to model the human body and track essential functions in real time, leading to more individualized treatment regimens and advancements in medicine. They also come in handy in urban and city planning and management; smart cities and management of and control of energy grids, traffic and transportation, and other facets, leading to better, more efficient, and sustainable living environments. Digital twins are not simply another application of IoT they are an essential tool for companies that want to generate future earnings in a world that is progressively extra logical and interconnected. This can be gauged in their rising adoption and documented necessity in specific applications.

Digital twins are becoming relevant in the manufacturing process in an ever-increasing fashion. These allow for real-time monitoring and predictive maintenance through model digital clips of manufacturers' manufacturing lines and machines. This skill reduces the time spent and costs associated with stoppages and overhauls and, in the long run, reduces the amount of money that must be spent.

Manufacturing intelligence results point out inimical trends, which makes it easier for manufacturers to improve the quality of the final product as well as the efficiency of the manufacturing process. Another field that has not remained untouched by the potential of applying digital twins is the healthcare sector. In this area, digital twins help optimize patient planning, obtaining virtual patient forms, thus ensuring that patients are given better treatment plans and they in turn get better. They can check or monitor their functions, mimic and monitor the human body's lower level, and anticipate possible complications. It has been notably useful in long-term care, in the development of new medications, and in surgery planning which makes it very fundamental in enhancing the overall quality of the services offered in the health facilities.

Significant achievements of digital twin technology have also been registered in urban planning and the Internet of Things (IoT). Digital twins are used for modelling and managing complex urban processes, like energy and other networks, structures of buildings, and transport systems introduced as smart cities. City planners can make decisions based on analysing the effectiveness of the current and new systems, making the urban area more sustainable, using less energy, and enhancing the quality of life of those living in the city by creating replicas of such systems. Another area in which digital twins are also applied is the Internet of Things (IoT) industry to achieve improvement in the performance of the connected devices and networks, as well as to warrant smooth functioning and communication.

2.3 Problems Related to the Adoption of Digital Twins

Besides, it is essential to underline that although the scenario of applying a digital twin in various kinds of businesses looks very promising, there are several challenges implied. Each of these difficulties includes numerous important sections, namely, technology challenges, privacy and security issues, and data credibility. In this article, the author intends to explain these concerns and pave out useful recommendations on how to ensure the proper implementation of digital twin technology. [1].

2.3.1 Challenges with Data Security and Privacy

Privacy and data security are major concerns when it comes to using digital twins. The risk of data breaches and privacy violations rises as organizations collect and use massive volumes of data to build and maintain digital twins. To tackle this difficulty, it is essential to:

- Use Robust Data Encryption: Some of the ways that information can be encrypted while transferring, and when it is stored is by ensuring that it is integrated so that nobody can easily access the data.
- Role-based permissions and access restrictions: Another measure for ensuring privacy is to regulate the input/output options that allow interaction with digital twin data for only those with access privileges. To control and limit access to the data accumulated it is necessary to limit the permissions based on roles.
- Adherence to Data Regulations: To the sector, complete knowledge of GDPR and HIPAA along with other laws and regulations is required. In this case, data anonymization and data pseudonymization techniques could also provide help.
- Train and Educate Staff: To reduce the possibility that human error will result in security breaches, train and educate staff members on data security best practices.
- Query Quality Issues: When implementing digital twins, data quality is crucial. The precision and dependability of digital twins can be compromised by poor data quality, which can impact prediction and decision-making capacities. To address issues with data quality, think about
- Data Validation and Cleaning: Perform various techniques to identify and resolve the problems and inconsistencies in the data. Data consistency can be achieved on the same subject with the help of a digital twin to remove any inconsistency that may be in the data used for the model.
- Sensor calibration: in case of Sensor Fusion errors and discrepancies noticed; always Calibrate Sensors & Data Source.
- Data Governance: In this context, your organization needs to have proper data governance because it will utilize data for various purposes and in different contexts throughout the data's life cycle.
- Constant Monitoring: Ordinary reporting systems and never-ending scrutiny should be established to detect such problems in data quality as soon as possible and address them [2].

2.3.2 Technical Difficulties

- Two typical technical problems related to the digital twin involve integration complexity and scalability. Experts noted that enlarging the digital twin concept and adopting data from different platforms can be a complex process.
- Standardised Data Formats: Where possible, promote the use of unified data-interoperability structures such as format compatibility and protocol compatibility
- APIs and Middleware: Data conversion software Middleware and Application callable interfaces are used for data integration and the exchange of data between two or more systems.
- Cloud-Based Solutions: cloud-pro professional solutions are scalable by essence, as are the variable resources and storage solutions that they integrate.

- Machine Learning and AI: To Facilitate integration of data and outline scalability possibilities for the future, and further business opportunities leveraging AI and machine learning

2.4 Challenges with Data Security and Privacy

Digital twins have brought tremendous opportunity and innovation in this era of digitalization for companies across industries. Digital twins are fully detailed replicas of actual tangible or intangible assets or processes that provide information and data exchange between the physical and virtual context and can also make predictions about the functioning of a particular object. In the present course of globalization and social reform, concern for data and information security and privacy has emerged as a paramount issue. Issues to do with data security and privacy concerning digital twins are highlighted in this section and, it establishes how important these issues are in preserving the dignity of the new kind of technology [2].

2.4.1 The Value of Privacy and Data Security in the Framework of Digital Twins

The essence of the digital twin concept is a set of data, as it is a continuous data stream that drives digital replicas of actual assets. That is the work that has been carried out with the help of Big Data technology, and this data includes all the sensitive information, trade secrets, and valuable operational insights; at the same time, this data is attractive for cybercriminals who want to steal it and commit security breaches. In this regard, it is meaningless to overstate the significance of data protection and privacy considerations when it comes to multisensory digital twins since data are a fundamental building block of these models.

The most obvious reason for data security is to neutralize threats that active digital twin systems face from hackers and hostile forces. The consequence could be staggering in that confidentiality is compromised right from the leakage of intellectual capital to posing a potential threat to the physical infrastructure. Firstly, access control measures are required to minimize the exposure of the collected data to the people who should not access them Secondly, while transferring the data or using them in various activities advanced measures like data encryption techniques must be used to enhance the protection of data, they should also follow laws like the Health Insurance Portability and Accountability Act (HIPAA) and the General Data Protection Regulation (GDPR). Protecting data privacy is essential, from both legal standpoints. Information in a twin environment typically includes personal, asset, and operational information. The mere fact that the privacy laws and regulations are complied with does not establish trust among stakeholders. Moreover, it is important to comply with the law as well. To implement privacy protection but still enable the use of the data in analytics, data anonymization and pseudonymization methods may be warranted.

2.5 Data Security Challenges

The digital twin begets a different set of challenges related to data security. All of them are linked to the possibility of unauthorized access and consequent breaches of data. In this section, the following specific security risks will be included—elaborated on how exactly complex these threats are and what kind of damages can result from these threats.

1. Access by Unauthorized Persons:

   One of the continuous problems is having unauthorized access to digital twin data. Digital twins capture essential data regarding operational procedures, systems, and physical assets. This hence means that in case the data finds its way to opportunistic persons or other entities, then this may mean disaster, financial losses, disruptions to business, and security and safety compromise for the assets. Challenges in regard to unauthorized access are numerous:

   - Weak Authentication: Systems with nil or weak authentication are readily accessed. Organizations shall put in place multi-factor authentication for the authentication of users to have access to the digital twin data.
   - Insider Threats: The employees and insiders accessing digital twin data are potential threats to the data. Organizations shall establish procedures to detect and prevent insider threats.
   - Lack of access controls: In the absence of access controls, it is possible to access data outside the granted permissions. You need strict access controls and role-based permissions to give people access solely based on the need to know.

2. Data Violations:

   Big risks to digital twin installations come from data breaches. A data breach can release private info, alter or delete without permission, and be catastrophic to finances, legal, and reputation. Among the challenges of data breaches are:
   - applications and System Vulnerabilities: This paper identified several common uses and systems that are usually associated with using digital twins. These elements can be exploited by antagonistic actors who control and manipulate them. To address this, security audit and constant updates must be a must-do solution to close that possible security weakness.
   - Phishing and Social Engineering: Phishing and social engineering are some of the most prevalent and often employed techniques used by cyber criminals to defraud people. Health programs aimed at sensitizing its employees and training them can help protect workers against such attacks.

- Data Encryption: If the data that is being transmitted and stored is sensitive, it can be exposed due to subpar data encryption. No data is compromised as the information is encrypted with highly modern algorithms to the extent of even with a data breach attempted.
- Data Resilience and Backup: It was clearly stated that the impact of loss caused by a breach can become minimal through backup measures and robust data resilience. In case of a breach, this assures that the data can be recovered to thwart the attacker's attempts.

It is therefore important that enterprises remain vigilant due to the constantly evolving nature of digital twin security and the risk of unauthorized access and data leakage. There are multiple security concerns regarding numerous digital twin applications; however, if organizations follow proper guidelines, perform regular check-ups, and always cover all preventive measures, then data security is more likely to be maintained.

2.5.1 Challenges in Data Privacy

Since digital twins often involve purchasing and utilizing data collection from multiple sources, privacy concerns are an integral aspect of their application. These problems revolve around the need to safeguard the individual liberties of a person whose information may be processed and other forms of confidential information. In this part, we will focus on the privacy implementation of the acquisition and use of data for digital twins [3].

1. Assent and Knowledgeable Data Gathering:

   The most important privacy concerns of digital twins are associated with the task of obtaining informed consent for data collection. Collectors of raw data for building or updating digital twins, therefore, have to guarantee the complete voluntariness of the use of personal data. However, getting the subjects' permission is not only ethical but is often necessary, following the standards of data protection laws. Obstacles in this field consist of: Obstacles in this field consist of:

   - Transparency: Letting people know or informing them in a manner that they easily understand what information will be taken, how it will be used, and who will be given access to the information awoke them.
   - Approval mechanisms: Adopting integrated procedures that permit easy access to prompt consent from the data subjects including simple opt-out techniques and opt-in options.

2. Minimising Data:

   Similar to data minimization and other privacy principles, there is the necessity of collecting only the information that is needed for the purpose. In a nutshell, the greatest challenge lies in determining exactly how much data is needed to create accurate digital environments without falling into the trap of collecting too much data to infringe on someone's right to privacy. This task consists of:

It is equally important to be clear on which data of the asset is relevant for the particular digital twin process Since one wants to avoid the accumulation of unnecessary information.

3. Pseudonymization and anonymization:

   In the case of the digital twin, data minimization approaches such as anonymization and pseudonymization are commonly used to protect the privacy of individuals. Obstacles in this field consist of: Obstacles in this field consist of:

   Data Utility vs. Privacy: Balancing can be defined as the process of searching for the optimal proportion that provides the best level of privacy to the individuals whose data is being processed by simultaneously maximizing the usefulness of data for operations and analysis by avoiding its anonymization or pseudonymization.

   Re-identification Risks: Knowing that such an option may exist and preventing this from happening, it is possible to protect one's individuality while working with data that seems to be anonymous.

4. Transnational Data Transmission:

   Or where the subject matter is in multinational corporations or sectors, cross-border data transfer may be a privacy concern. This entails ensuring that whenever data is transferred across jurisdictions, it is safeguarded and meets the legal requirements of the relevant jurisdiction.

   Data Transfer Agreements: Ensuring compliance with data protection requires assessment of data protection regulations in different particulars and encouraging data transfer agreements.

   These matters need to be handled with certain dynamism relative to the current state of data privacy regulation as required under CCPA, GDPR, and any other laws enacted to ensure ethical data management. In the concept of digital twins, managing the relationship between the creation of meaningful digital twins and maintaining people's privacy rights is an ongoing and fluid challenge. As for those privacy issues to which we have not provided a direct solution in this section, we will suggest ways of dealing with them in the next section.

2.5.1.1 Examples of Security and Privacy Issues in Digital Twin Projects in the Real World

Breach of Smart City Infrastructure Security

- Challenge: Unauthorised Entry
  There is a sad story of an incidence of unauthorized access in one of the smart city digital twin projects that brought up several critical security questions. In this attack, there was a vulnerability in a third-party program that led to unauthorized entry into the digital twin system, which contains features of the munic-

ipal infrastructure. As a result of the hack vital details about electricity systems, public buildings, and transport networks were uncovered. The attacker was able to modify real-time data because they had access which they were not supposed to; this affected traffic signals and deceived inhabitants of the city [4].

- Resolution: Strengthening Access Control
  After the intrusion was realized, the local officials enhanced the access control mechanisms and installed protocols for strong authentication and permission. Further, they performed a security review where they looked for, and solved, vulnerabilities in the digital twin environment. This event shows that the issues of protection of digital twin data should be solved with the help of strict access controls.

Leakage of Digital Twin Data in Healthcare

Challenge: Data Privacy Issues
An example of a data privacy event occurred in a renowned healthcare digital twin pilot project for creating detailed patient profiles. An email of patient data was sent from an employee by mistake to another person as it was not encrypted. Such a situation poses a lot of concerns, particularly to patient privacy and regulatory compliance in healthcare organizations.

Resolution: Training and data encryption are the two types of training that can be discreetly incorporated into healthcare settings.

It, in turn, broadened the approach the healthcare uses to protect patient data in transit, minimizing the risk of its being hacked. They also provided institutional memory and a huge dose of training for the staff members in order to reduce the likelihood of human error when dealing with data. Consequently, this event was pretty much of a wake-up call regarding data security measures while patient information can be part of healthcare digital twin solutions.

Data Breach at a Manufacturing Plant: Data Breach at a Manufacturing Plant

Challenge: Data Breaches and Unauthorized Access
This is an example of a real cyber intrusion that resulted in leakage from the digital twin system in such a manufacturing digital twin project. Information concerning the manufacturing process, such as the production calendar and designs of the models, was leaked, they noted. This simply hacked the company's competitive edge and put into question whether ideas constituting part of the business model were safe.

Resolution: Improvement in Safety Features
For this reason, the manufacturing company increased its access control and added more layers to its encryption system among other security measures. They have also initiated monitoring of any suspicious abnormal activity that may be happening in their twin digital environments. That specific incident served to put a spotlight on

the need for protecting confidential information about productions and proprietary rights in smart twins [4].

1. Proposals on How to Ensure Data Security and Privacy within the Application of Digital Twins

   Digital twin technology is such a fast-developing industry; as a result, organizations have to protect the security of data concerning privacy protection. The following recommendations would help ensure security of data and privacy protection in deploying digital twins [3]:1. Reliable data-encryption:

   Proper security processes, such as data encryption across in-transit and at-rest, would make it safe to protect confidentiality. Protection of data at the instance of transfer amongst these components, and within the framework of the digital twin itself, should always be ensured.

   Encode your data to make sure that the embedding of the features of privacy is perfect through the use of better encryption algorithms like AES-256.

   It is necessary that the techniques of encryption are updated with the highest frequency possible to address new types of threats and vulnerabilities.
2. Role-based permissions and access restrictions:

   If you want to control who can access the data, then you have to take some strong access control measures. To ensure that users only search for information that is relevant to their work, use role-based permissions.

   Other social controls include monitoring users and auditing to detect any unauthorized or suspicious activities.

   Despite the improvements made, the way you access your car is still insecure, especially for users with a high level of rights applying multi-factor authentication.
3. Respect for Applicable Regulations:

   Always read and follow up on any laws that are relevant to your space or company—be it CCPA, HIPAA, or GDPR.

   Establish a compliance/compliance team or data protection officer who will be responsible for ensuring compliance with a particular legislation or regulation.

   For the identification and prevention of privacy issues, data protection should be conducted on an impact assessment test.
4. Minimising Data:

   This is a reminder that data-collecting processes should be checked often to ensure that none of them is collecting unnecessary information.

   If the information gathered is sensitive, one should try to make the data anonymous or pseudonymous to retain its functionality while minimizing the invasion of privacy.
5. Reduced Insider Threat:

   Outlines elaborate procedures on how to detect and control insider risks. Promote a security awareness culture among the staff to ensure they are aware of potential attacks from insiders.

Be alert for strange or suspicious patterns in the use of the system, particularly, by system administrators or any other users with the right to access special files.

Ensure that a restricted number of organizational members have the privilege to access large volumes of information via clear and strict access control measures.

6. Constant Security Evaluations:

    Ensure Security Ensure regular security auditing and vulnerability assessments are conducted to ensure that any flaws in the digital twin system are identified and corrected.

    Perform intrusion detection and prevention by monitoring the instances of security breaches and unauthorized access through technology.

    Keep yourself updated on the new threats and vulnerabilities and update your security accordingly.

7. Data Backup and Resilience:

    Make backup plans and data resiliency so that in case of breach or loss of data are restored immediately.

    Design disaster recovery plans, including data restore procedures, including that of the digital twin.

8. Training on Security Awareness:

    Provide staff members training on privacy and data security practices.

9. Event Reaction Strategy:

    Establish and maintain the incident-response plan- documenting the steps to take should a privacy or security incident occur-. Use simulation exercises to test the crisis response strategy and be sure that the organization is ready for the implementation of an appropriate response.

    These guidelines, if implemented, can greatly improve digital twin deployments concerning the security and privacy of data. Associated risks will be mitigated; ethical and responsible usage will be ensured. All these steps must be taken to engender stakeholder trust and ensure the long-term sustainability of digital twin initiatives in many sectors.

2.6 Data Quality Challenges

Good quality data is at the foundation of digital twin applications. It ensures that virtual replicas of physical assets or systems are accurate, reliable, and effective. Whenever writing about digital twins, one will very often stumble upon the words "completeness, accuracy, consistency, and timeliness", which refer to how the data is used to create and sustain digital twins. In this section, we will talk about high-quality data's integral role in digital twin applications [4].

High-quality data is a priority for digital twin applications.

Precise Illustration
There's an idea behind each digital twin: it's copying physical reality digitally effectively and accurately. The proper representation of its physical version by a digital twin requires high-quality data. If that data becomes inaccurate or incomplete, the utility of a digital twin suffers.

Trustworthy Decision-Making
It is in simulations, predictive analysis, and decision support that digital twins are used very heavily. The results and predictions arrived at in the digital twin are improved when the quality of data is better. With increased confidence in the fact that data driving these decisions is valid and accurate, organizations confidently make some decisions on resource allocation, process optimization, and maintenance. It also helps give expression to, analyze, and model data along the lines of energy consumption, environmental issues, and transportation systems by urban planners in the instance of smart cities, leading to more sustainable and efficient urban development.

Functional Effectiveness
Digital twin differential deployment for operational effectiveness It is a good time to talk about what digital twins bring to operational effectiveness. Good data helps: find inefficiencies, reduce downtime, and optimize manufacturing processes-logistics-health care, or any other process- for effective outcomes The digital twins help get accurate information on the status and performance of equipment, items in a supply chain, or a patient's condition, which the manufacturing facility uses to predict maintenance, thereby lowering the cost and occurrence of maintenance and downtime. It is realized by obtaining real-time data of the twin in the supply chain through the control of routes and inventory in logistics, thus cost reduction, and improved customer experience.

Innovation Driven by Data
High-quality data formulates the basis of data-driven innovation. If organizations use digital twins that are constructed based on dependable and accurate data, then they can develop and explore new avenues. High-quality data, for instance, helps engineers and designers to create better products in the product designing and development process by modelling the performance precisely and spotting the areas where improvements are needed. High-quality patient data fuels personalized treatment plans and ground-breaking research in healthcare.

Risk Reduction
Poor data quality can serve as a source of vulnerability and hazard. Good quality data can be safely used for modelling security risks and applications of vulnerabilities, including digital twin-based cybersecurity, where digital twins are created to simulate and neutralize potential threats against it. Poor data might lead to some risk being unaccounted for or some security measures omitted.

2.7 Sources of Problems with Data Quality

Poor data quality can disturb performance and may have an impact on the precision, dependability, and usability of the data utilized in digital twin applications as shown in Fig. 2.1. It is essential to locate and fix these causes of poor data quality to guarantee that the digital twin performs as planned. Here are a few typical causes of problems with data quality:

1. Errors in the sensors:

   Many applications that use digital twins rely on data from sensors. Some failures of data quality can be traced back to the making of incorrect measurements by such sensors. Damaged or malfunctioning sensors can deliver erroneous and sometimes untrustworthy data; drift and calibration effects can also distort data, which impairs the accuracy of the digital twin to reflect the real system.

2. Anomalies in Data:

   Data transmission errors, failure of the devices, and environmental conditions are just but a few causes of data anomalies. Such inconsistencies may bring noise into the data so much that sometimes it becomes hard to differentiate between irregularities and occurrences that happen in the real world.

3. Issues with Data Synchronisation:

   Sync issues may appear when the digital twin relies on data that originates in a few different systems or sources. The inaccuracy of aligning the data in time may produce differences in the real-system representation by the digital twin. Such problems lead to inaccurate predictive models and simulations.

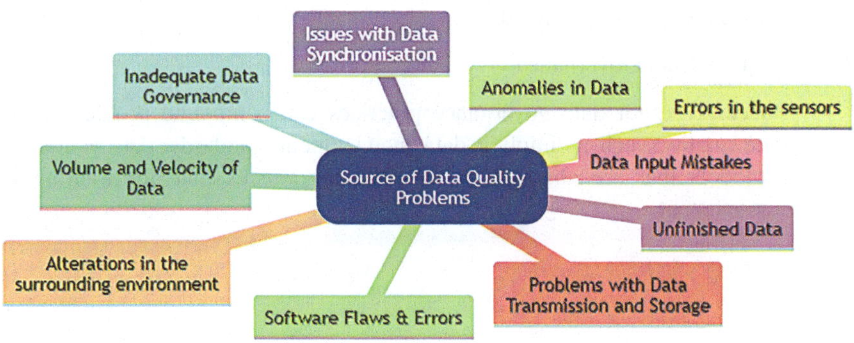

Fig. 2.1 Problems with data quality

4. Unfinished Data:

    The data can be incomplete because some points missing the Process data, or not all relevant factors have been noted. The digital twin would then just not capture the entire and correct state of the physical system.

5. Data Input Mistakes:

    Inaccurate and inconsistent data can simply be due to human error in data entry or migration. Such inaccuracy can be as simple as typing errors, wrong units of measurement, or even wrong interpretation of the data sources.

6. Problems with Data Transmission and Storage:

    The quality can be lost in transmission and storage. Corrupted files, loss in transfer, or failure in backup or recovery processes can result in data used in the digital twin being unreliable.

7. Alterations in the surrounding environment

    Data loses their accuracy under some conditions, depending on environmental factors in the application. For example, where there is influence after humidity, temperature, or other external factors, they may likely affect the accuracy of sensors or the reliability of data transmission.

8. Software Flaws & Errors:

    Software that deals with data, either to store it or process it for checkout in a digital twin, may have faults that reduce the quality of the data. Others may lead to improper interpretation of data, or even to develop wrong models and simulations.

9. Volume and Velocity of Data:

    High data velocity and volume to be processed may in some applications present an amount of challenge. It means that systems tailored towards efficiently handling very large volumes of data could otherwise see a deterioration in the quality of such data.

10. Inadequate Data Governance

    Successive poor data governance practices can compound problems with data quality, such as implicit data management standards, data validation protocols that are absent, and data quality checks that are wanting or non-existent.

2.8 Impact on Making Decisions

Poor quality of data might lead to far-reaching and profound consequences related to the operations and decisions of an organization. Digital twin applications, which greatly rely on quality and reliable data, are more susceptible to the adverse impact of poor data quality issues. Some ways through which poor quality of data might affect operations and decisions include:

1. Erroneous Decision-Making

   If the digital twin operates on low-quality data, then the insights and information you acquire from the digital twin are going to be defective. Hence, the decision-makers will make less informed decisions mainly because the conclusions that they reach will be based on information that is inaccurate or incomplete. For instance, in a manufacturing digital twin, inaccurate data from sensors may mean giving the wrong maintenance recommendations on the equipment, which triggers a domino effect that could lead to gratuitous downtime or unnecessary repairs.

2. Decreased Self-Belief in Making Decisions:

   Poor data quality means decision-makers lose confidence in the insights the digital twin is providing. Data that could be inaccurate leads to a lack of belief in how reliable that system is, finally causing decisions to be delayed, scrutinized more, and a reservation to accept changes to improve.

3. Inefficiencies within operations:

   Bad data quality can impede operational optimization. Incomplete or inaccurate data will lead to inefficient use of resources, inefficient processes, and lost possibilities for advancement. For instance, faulty demand forecasting due to poor data in a digital twin in the supply chain management sector will lead to overstocked or under-inventoried situations.

4. Enhanced Failure Risk:

   Predictive maintenance is mostly done by building digital twins so that equipment breakdowns can be stopped. On the contrary, data of poor-quality results in prediction models fail to capture early indicators of equipment breakdown or failure. This results in costly repairs, unforeseen breakdowns, and the possibility of safety implications.

5. Impact on Customers:

   It means that in things like healthcare and smart cities, those areas where digital twins are used to bring new experiences to end-consumers, the impact of poor data quality will impact the resulting services. In the case of the healthcare sector, for instance, if a patient-specific digital twin is not so accurate, treatment suggestions may be inappropriate and that could put the patient's wellbeing and even trust in the professional at risk.

6. Waste of Resources:

    There always is a payment in terms of correction of errors, review of decisions and judgments, and decrease of business disruption to deal with poor data quality aftermath. This translates into financial, labor, and time resources expended that could be better used elsewhere.

7. Behavioural Harm:

    Poor quality of data can hurt operations and decision-making that are beyond corporate boundaries. Loss of confidence and consequently the organization's reputation when consumers, clients, or stakeholders feel insufficient expertise in data management and decision-making exists.

8. Passed Up Chances:

    Poor data quality will bury important trends, opportunities, or valuable insights. Accuracies and omissions in the data could cause organizations to miss an opportunity to earn a competitive advantage, customer satisfaction, and optimization of business processes.

2.8.1 Techniques for Increasing the Calibre of Data

Improving the quality of data directly applies to the accuracy, reliability, and speed of the digital twin applications; therefore, improving data quality is crucial to these applications. These are some doable tactics to improve the quality of your data: These are some doable tactics to improve the quality of your data:

1. Validation of Data:

    To eliminate errors and inconsistencies, use the data validation tests in some smaller steps of the data gathering and processing. Some of these checks include tests such as outlier detection, format validation, and range validation to mention just a few. Again, when creating digital twins, ensure that values from the sensor, for instance, are within the required or expected thresholds, and mark or modify those that deviate much from the typical values.
    Statistical techniques can help identify outliers and anomalies in data. These techniques, along with box plots and Z-scores, will also lead you farther from the mean data points to find possible errors.

2. Data Purification:

    Data cleansing is the process of detecting and correcting such errors and inconsistencies. This can be in the form of typographical errors, removing duplicates of records, and also filling up missing or incomplete data, etc. For instance, data cleansing for a healthcare digital twin can correct patient records with incorrect demographic data or medical history information. Automate data cleansing procedures to gain productivity. Most common problems in data quality can be identified using data profiling tools, and data points that are problematic can automatically be corrected or flagged using data cleansing software.

3. Sensor Adjustment:

   Calibrate sensors regularly to provide accurate readings. Calibration of sensors implies that one compares the measurements made by the sensor with a known reference and applies the appropriate corrections. In applications where high confidence in the data derived from sensors is necessary, for example in industrial processes or even environmental monitoring, calibration of sensors will be even more critical. Apply standard methods of calibration, and always document the activities of calibration. In this way, it is always ensured that the data from the sensors is accurate and reliable.

4. Data Management:

   Clearly articulate the guidelines and processes of data governance in your organization. Define the rules, roles, and responsibilities for data stewards, and data quality teams among others. Articulate the guidelines and standards of data quality that inform the acquisition, validation, and cleansing of data. The guidelines should align to specific requirements of the applications that use digital twins.

2.8.2 Integrity Difficulties

As part of their implementation, digital twin initiatives are sometimes faced with problems arising from the integration of data from various sources and systems. The need for harmonization and convergence of data on a single platform for seamless digital twin representation, and data source heterogeneity develop this form of complication. Integration problems include:

- Data Format Divergence: Source data can be in varying formats and according to different standards; hence, the integration exercise will be thus required to map and transform these disparate data formats into a standard structure that is relevant to the digital twin.
- Interoperability Challenges: Integrate systems and devices normally not speaking to each other and is one of the first common steps in data integration. The lack of interoperability standards may cause compatibility problems forcing an organization to build specific interfaces for or middleware.
- Real-Time Data Sync: There is a challenge in genuinely actualizing real-time synchronization of data from multiple sources. The digital twin's timeliness and

accuracy may be influenced by delays or inconsistencies brought about in updating data in applications that require updated information.

- Data Volume Management: Numerous sources processing huge amounts of data can overload processing, storage, and network resources. The scalability of infrastructure and proper techniques of managing data shall enable the crossing of these hurdles.
- Consistency and Quality: While integrating data from multiple sources, the quality of data is likely to be erroneous and inconsistent. Quality can be maintained in the integrated data set only by cleansing and validating data.

2.8.3 Scalability Problems

Demands on scalability—particularly related to data volumes and usage cases—are common in many digital twin applications. Such demands can be related to resource allocation, system performance, and cost-effectiveness.

Resource Scaling: The faster the digital twin ecosystem is adopted, the more it becomes very clear that network bandwidth, storage space, and processing power have to also be put in place to cope with the scaling of data volumes in data processing.

- Latency Management: In such applications as the Internet of Things and smart cities, scalability shall not be achieved at the cost of real-time or low-latency needs Scalability demands increased handling in terms of users, and traffic, but that should not lead to appreciable delays.
- Cost Control: Increased infrastructure also tends to increase operation costs. Therefore, a need to save on operational costs by embracing efficient cost management practices like adopting cloud-based solutions, which mount sparingly on resources and, in the end, effective data storage.
- Flexibility of Architecture: Scalable digital twin architecture should adapt to the changing needs. It should be feasible to add new sensors, other devices, and data sources without reconfiguration.
- networked Systems: Interoperability issues may also occur because scaling may make facilitated devices and systems that are networked decrement interconnection. There needs to be solid management and maintenance of the links for this scalability to be as seamless as possible.

Flexibility and adaptability are key to successful digital twin implementations while handling continuously changing difficulties on integration challenges and scalability issues, which need careful planning and design of infrastructure and continuous monitoring.

2.9 Technical Difficulties in the Development and Maintenance of Digital Twins

There are several technological challenges concerning the design and maintenance of digital twins, which need to be surmounted if fulfilling use is to be made of this game-changing technology. Digital twins prove to be effective in the case of modelling complex systems extremely well. This holds in the case of industries like manufacturing, smart cities, healthcare, and more. However, their effectiveness depends upon surmounting technological challenges relating to scalability, integration, and complex system modelling. In this section, some of the related technological challenges in developing and maintaining digital twins will be reviewed, ensuring the potential uses of such systems for ever-increasingly complex systems [5].

Technical Difficulties

- Complexity of Integration: Data that is coming from many sources, like sensors, devices—even the Internet of Things to older systems—is used very often for digital twins. The hard part is bringing all this divergent data into one single, unified model. Different formats of data, interoperability issues, and a call for data integration in real-time make integration quite difficult.
- Data Security and Privacy: With the huge amount of data that digital twins collect, the greater the challenge of ensuring safety from breaches and other unauthorized people. Privacy issues further complicate data security processes. These also require compliance with pertinent laws on data protection.
- Scalability Issues: Because digital twins are deployed in dynamic contexts unlike static, where changes happen over time, systems have to be designed to support larger volumes of data, increased processing but also more sensors or devices. Scarce are the conditions guaranteeing scalability and, at the same time, low latency and low cost.
- complicated System Modelling: One of the major advantages of the digital twins is that they can completely simulate complex systems. However, because of its complexity, it is quite difficult for the digital twin model to capture complex relationships, feedback loops, and interactions.
- Real-Time Data Management: In many cases, it may be hard to synchronize real-time data, particularly in cases of the Internet of Things and smart cities. Inaccurate or delayed updates of data may diminish the purpose of using a digital twin.
- Resource Allocation: Careful allocation of resources is needed for scaling the digital twin infrastructure: computing power, space for storage, and network bandwidth. Without proper management of those resources, one cannot ensure system performance while limiting expenses at the same time.
- Flexibility of Architecture: The architecture of digital twins must be flexible to change as per requirement. It must be easy and quick to add new sensors, devices, or data sources; its addition shall not make a lot of reconfiguration or otherwise illustrate.

- networked Systems: Interoperability issues make the maintenance of networked devices and systems challenging and will find solutions for a seamless operation just as digital twin ecosystems are now developing.

Organizations have to, therefore innovate in this dynamic digital twin technology to meet these technological challenges. Successful working of all those complications is required to establish a digital twin in systems that are increasingly complex and in a range of industries.

2.10 New Technologies and Industry Best Practises to Address Technical Difficulties in Digital Twin

Best practices and emerging technology enable organizations to surmount the technical challenges of digital twin projects. The tactics, in general, improve overall digital twin success and relate to integration, scalability, and complex system modelling as well [5–7].

1. Edge computing: new technologies, such as Edge computing, work on the very basis of latency reduction and network bottleneck relief by processing the data closer to the source of the data. It is very effective in the case of digital twins for real-time synchronization of the data.

   Best Practice: Analyze and filter this data locally and install edge computing nodes at sensor or device sites, reducing latency and Saving volume data effects.
2. Artificial Intelligence and Machine Learning: AI and ML could be used to make the predictions of a digital twin more accurate and powerful. These are the technologies that are in a position to make real-time decisions and learn from the previous data.

   Best Practice: To that end, a good practice is to include algorithms of prediction maintenance, anomaly detection, and data validation to avoid issues that look like Machine Learning. All those are tools that shall help find the problem with data quality and make processes less painful.
3. Blockchain Technology for Data Security: Internally, it creates a data ledger that is decentralized, invulnerable to breach, with strong privacy protection and data security safeguards.

   Best Practice: That digital twin data is maintained private, unchangeable, and secured using blockchain technology. The issue of data security vast and unwanted access can be enhanced by the introduction of blockchain technology.
4. Emerging Technology and Digital Twin Standards: Development that could enable integration and ensure compatibility of various components of digital twins is through the development of interoperability protocols as well as standards.

Best Practice: Use still developing digital twin standards, for instance, ISO 23247, to uniformly and in a coordinated way develop digital twins to overcome integration and interoperability issues.

5. Augmented reality and advanced visualization: Emerging technologies offer the promise of a better way of representing complex systems through advanced visualization techniques and AR digital twins.

   Best Practice: The technologies can be used to develop dynamic and immersive digital twin worlds through augmented reality and advanced visualization techniques. This will enhance an understanding of the major, intricate system models.
6. Best Practices for IoT Ecosystem Management: Organize IoT ecosystems per best practices, including the proper management of devices, firmware updates, and security measures that mitigate concerns over data security and privacy.
7. Quality Control and Data Governance: Establish strong data governance practices, for example, data quality validation; data cleaning needs to be done to uphold the quality. The organization needs to actualize quality control teams and data stewardship to ensure high quality.
8. Constant Analytics and Monitoring: Recommended Practise: Monitor the security and performance of digital twin systems at all times. Advanced analytics should be applied to the digital twin system to proactively identify issues related to data anomalies or possible breaches.

Such complexities and challenges can be managed if organizations embrace this futuristic tool and other best practices in digital twin projects. These strategies enhance data quality, data protection, and system run time so that the existing and newly created digital twins can model real-world systems more effectively and aid in better decision-making, a central aim of operational excellence across various domains.

2.11 A Comprehensive Method for Ensuring Data Security, Privacy, and Quality

A multidimensional approach is needed to ensure data quality, security, and privacy in digital twin initiatives. Here is a comprehensive plan to deal with these important factors [8, 9]:

1. Governance of Data:

   Ensure clarity of responsibilities for data management by aiming at getting sound frameworks for privacy, security, and data quality. To oversee these areas, hire specific teams focusing on data quality and proper stewardship.
2. Data Quality: To increase the quality of the data, perform data validation and cleansing methods in an attempt to search and eliminate errors in the data set. Utilize data purification software, data auditing software, data differentiation,

and statistics utilization. For the monitoring of the input data accuracy, punctuality, and thoroughness in the model of the digital twin, conduct regular auditing and reviewing of the data.

3. Data Security: Data encryption is another security measure that should be implemented to protect data even in the middle of the communication and when they are stored. Employ stringent encryption standards and ensure to learn about new encryption standards appropriately.

   Restrict user to view data by granting user accesses according to the role assigned to them. In an effort to detect any possible intrusion, misuse or unusual activity, it is also imperative to monitor and review the access Control.

   For checking the weaknesses and problems in the digital twin system and to prevent hacking, security audit and vulnerability assessment. Implement IDPS for monitoring and detecting unauthorized access to your computer system
4. Data Privacy: Ensure the users provide their consent before their data is collected and ensure that the users have prior knowledge of when they can have an access to their data. Ensure that data subjects have choices in the process; namely, choices to participate in the processing of their personal data and to withdraw their consent from the processing of their personal data.

   When collecting and storing such data, one common best practice is to anonymize or pseudonymize the data so that the usefulness of the data is maintained, while the privacy of its subjects is protected.

   Transfer personal data only within secure networks and review the local legislation concerning data transfers, wherever applicable.
5. IoT Ecosystem Management: Ensure that the IoT assets have security measures and the firmware is updated. Ensure the security of the connections and devices is constant and updated.
6. Blockchain for Data Security: Consider applying blockchain as a means of guaranteeing data security in a distributed database that can be accessed only with a key. Thereby the protection of the data confidentiality and the data integrity could be enhanced.
7. AI and ML for Anomaly Detection: Implement machine learning and artificial intelligence to identify existing defects in real time and on scheduling maintenance beforehand. Use these tools to identify risks and vulnerabilities in the security area, and to check data quality issues.
8. Employee Training: Provide specialised guidelines on suggested practices for securing data to the staff members formulating policies. Make the people feel responsible and aware of their work and that of those around them.
9. Compliance: Comply with protection acts' sister standards; for instance, CCPA, HIPAA, GDPR, among many more regarding data protection legislation. Nominate the compliance team or Data Protection Officer who can handle the regulations at work in an organization in their compliance.
10. Continuously Monitor and Optimise: Observe how the digital twin system is working; whether it is secure or whether the data privacy is compiled with. Resolve problem areas and threats with proficient key performance indicators for the application of advanced analytics.

2.12 Conclusion

The adoption of digital twin technology now more than ever emerges very strongly as a definite catalyst to innovation, efficiency, and well-informed decision-making across many industries in today's digital age. The chapter proceeds further to take a deep dive into the controversies of digital twins by exploring the opportunities and problems associated with their creation and maintenance for an organization. Beyond data quality and quality control, a number of challenges range from purely technical difficulties to data security and privacy. The bottom line here is that such challenges cannot be lacking. Through the implementation of best practices, integration of new technologies, and diligent activity in relation to data governance, organizations will be able to face these listed difficulties. Advanced characteristics in data security and privacy involve encryption, access controls, and blockchain technology. Good data validation, cleansing, and calibration techniques improve its quality. In the past, further leverage of AI, ML, and edge computing has made the digital twin scalable, real-time capable to work with complex systems.

The potential of digital twins is virtually endless. Their applications inspire innovation, paving the way toward the future—right from smart cities to healthcare and manufacturing. Safeguarding the integrity and confidentiality of data, what fuels such systems, there is only one way to harness all that's powerful about digital twins—and that is through an end-to-end view on data quality and securityelfare. The chapter has provided an overview of the landscape, importance across industries, and problems and their solutions that set the basis for this ever-changing domain. Applying such principles of data quality, security, and privacy will help realize the full potential of what a digital twin can do in terms of data-driven decision-making, smooth operations, and building a more promising and connected future.

References

1. Menon D, Anand B, Chowdhary CL (2023) Digital twin: exploring the intersection of virtual and physical worlds. IEEE Access 11:75152–75172. https://doi.org/10.1109/ACCESS.2023.3294985
2. Attaran M, Celik BG (2023) Digital twin: benefits, use cases, challenges, and opportunities. Decis Anal J 6:100165
3. Botín-Sanabria DM, Mihaita A-S, Peimbert-García RE, Ramírez-Moreno MA, Ramírez-Mendoza RA, Lozoya-Santos J, d.J. (2022) Digital twin technology challenges and applications: a comprehensive review. Remote Sens 14:1335. https://doi.org/10.3390/rs14061335
4. Holmes D, Papathanasaki M, Maglaras L, Ferrag MA, Nepal S, Janicke H (2021) Digital Twins and cyber security – solution or challenge? 2021 6th South-East Europe design automation, computer engineering, computer networks and social media conference (SEEDA-CECNSM), Preveza, Greece, 2021, p 1–8, https://doi.org/10.1109/SEEDA-CECNSM53056.2021.9566277
5. Hua EY, Lazarova-Molnar S, Francis DP (2022) Validation of Digital Twins: challenges and opportunities. In: 2022 winter simulation conference (WSC), Singapore, 2022, p 2900–2911, https://doi.org/10.1109/WSC57314.2022.10015420

6. Khan LU, Han Z, Saad W, Hossain E, Guizani M, Hong CS (2022) Digital twin of wireless systems: overview, taxonomy, challenges, and opportunities. IEEE Commun Surv Tutor 24(4):2230–2254. https://doi.org/10.1109/COMST.2022.3198273
7. Rasheed OS, Kvamsdal T (2020) Digital twin: values, challenges and enablers from a modeling perspective. IEEE Access 8:21980–22012. https://doi.org/10.1109/ACCESS.2020.2970143
8. Mihai S et al (2022) Digital twins: a survey on enabling technologies, challenges, trends and future prospects. IEEE Commun Surv Tutor 24(4):2255–2291. https://doi.org/10.1109/COMST.2022.3208773
9. Fuller Z, Fan CD, Barlow C (2020) Digital twin: enabling technologies, challenges and open research. IEEE Access 8:108952–108971. https://doi.org/10.1109/ACCESS.2020.2998358

Chapter 3
Digital Twin Implementation: Architecture and Development

3.1 Introduction

The introduction/innovation toward the constructive design of the founders, which incorporated flexibility among virtual templates and physicality conveys them as one of a number now emerging digital twin technology positions. This is because functional, scalable, and adaptive concepts are no considered to first arise from in architectural design issues that act as templates for such efficacy. The chapter goes deep into the essence of architectural and design elements in terms of digital twins, uncovering only a delicate interaction between wanting out organizational objectives to be realized and being constrained by technology architecture [1].

As previously stated, the importance of designing a robust data architecture within digital twin systems is equal to other problems. Storage, processing and analytics of data from sensors, IoT devices and external sources have become more pertinent with the mushrooming variety on the expanded volume aided by a complex architecture. The findings of the work done by Wang et al. reveal that "a model data architecture acts as an analogous skeleton to digital twin systems enabling data integration and gain insights across domains". Information must range from real-time sensor data to historical trends incorporating broader categories of information necessary to remain on a global level [2].

Under interoperability, the importance of this quality comes out too clearly in digital twin architectures. As organizations digitize and embed digital twins in enterprises, the architecture has to be such as it enables irrespective communication capabilities among separate components. The standardized APIs, data formats and communication protocols serve as linchpins of interoperability or/and paramount collaboration both in distinct domains as well as holistic understanding. The industrial Internet Consortium (IIC) reaffirms interoperability and further says that "interoperability of digital twin system is a key prerequisite for successful cross-industrial integration". In addition, the design-based decisions have a direct correlation onto the scalability and optimization of digital twin systems. With the

© The Author(s), under exclusive license to Springer Nature Switzerland AG 2025

S. Gupta et al., *Digital Twins*, https://doi.org/10.1007/978-3-031-76564-3_3

implementation aspects of digital twin expected to increase in size of data sets and in introducing more user requests, it will be an important for architecture to possess attributes that support dynamic scaling. Scalability herein implies cohesive such as distributed computing, load balancing and responsive front end for the user. The highlight on "architectural scalability is key to the responsiveness provided by digital twin systems as complexity and scale conditions continue accumulating over time" [3, 4].

The chapter also underscores the inextricable link between architectural choices and security concerns. A well-designed architecture embeds security measures, including data encryption, access controls, and authentication mechanisms, to safeguard sensitive data traversing the digital-physical boundary. The Global Manufacturing Leadership Council (GMLC) highlights that "security by design is an essential principle that architects must embrace to fortify the resilience of digital twin systems". This chapter sets the stage for a comprehensive exploration of the intricate decisions that practitioners navigate in crafting digital twin architectures that empower organizations to bridge the gap between the virtual and physical worlds.

3.2 Constructing the Digital Twin Ecosystem

A digital twin represents the core nature of system architecture, which is an expansion in digital twin technology and integration. Organized by order and collaborates together as a system to create, work towards, and optimize digital twins. Something like waiting for the skeleton of a body to help support it in its journey into this world, the system architecture is used as a platform or structure that allows data collected from the real world seamlessly integrate with virtual simulations. The construct of this architectural masterpiece consists from several fundamental elements that it is thickly interwoven with [2, 5, 6]:

(i) Data Ingestion and Collection Layer: This lower layer of foundation provides an access, capturing data sensor networks, IoT devices and many sources that originate in the real world. It provides the essential link between what is physically happening and what is apparent in the digital twin world, providing real-time data awash.
(ii) Data Processing and Analytics Layer: The layers above this one go even further; more specifically, beyond the raw data that is there is the architecture of such an entity which makes it possible at this level to process as well as analyse the information. In this case, algorithms, machine learning models and analytical tools are combined together to generate patterns, trends and insights from raw data then provide valuable when making business decisions.
(iii) Virtualization and Modeling Layer: This is the center of this architecture which has where there digital twins will live their virtual life. By careful modelling and simulation this layer replicates the nature, physical parameters,behaviours

of their physical partners having thereby between them a dynamic virtual digital partner.

(iv) User Interface and Interaction Layer: Allowing user engagement is located at the middle layer, presenting interfaces and visualizations that grant users an insight into digital twin. This interface has user-friendly means of translating complex data into simple insights for effective decision making to the benefit of users.

(v) Communication and Integration Layer: Bridging each of the architectural pieces on a smooth manner is by means of what happens to be the Communication and Integration layer. This layer promotes easy data sharing and interconnectivity within the different systems, stakeholders, and various digital twin instances.

(vi) Security and Privacy Layer: The security and privacy architecture are designed to support the transmission of data between the worlds. This sort of shield includes encryption, access controls, authentication mechanisms and regulatory compliance.

Important Questions of Architecture The factor pins up the entire system architecture so that its effectiveness depends to a great extent on some key questions.

(a) Scalability: A comprehensive architecture that deals well with increased datas, data related operations and infrastructure complexity does not compromise speed or interactivity.
(b) Interoperability: The streamline of the integration of digital twins with other systems requires application design, and Programming interface aligned with defined standard communication protocols.
(c) Real-time Responsiveness: The live adaptivity in the real-time interactions between virtual and physical entities providing users with immediate effects of decisions on artefacts.
(d) Data Security and Privacy: Within every section of the data in transit and dedicated are additional security actions that people must be integrated to the architectural layer to protect safety with privacy legislation.
(e) User Experience: Convenient and intuitive user interfaces are concerned, which define the adoption rate by users and provide appropriate decision support in the digital twin system.
(f) Adaptability: An agile architecture is prepared for new technologies, dynamic demand changes and in challenges during disruption of industries, where the relevance remains unchallenged.

3.2.1 Real-World Implications

All such architectural decisions has a tendency to bounce through the apparent manifestations. Suppose a decentralized infrastructure adopted in the manufacturing plant. The use of edge devices for real-time processing on the production line in a plant environment that allows the realization of predictive maintenance minimizing production standstill time. Similarly, in a smart citywise infrastructure, a hybrid architecture is incorporated known as cloud resources are indispensable for simulations and edge devices permit real time monitoring while they enable put urban planning and resource allocation [7, 8].

Systems architecture is the articulation of the fabric in which reality and virtuality meet to integrate as digital twin technology. He carries out data carriage; energizes aetiologies, activates interactions to below fatherly roots behind biogenic powers at the bottom of digital twins, in ways that are present. In our chapter-based discussion, it is the nature of paradigms that are uncovered expose the inter fit pt. of architectural decision-making scope. The chapter on the finally, brings consumers close enough to enter into a world of digital twin to unravel its extremely aesthetic architectural imbroglio.

Data Ingestion and Integration Layer

As regards this compound symphonic nature of digital twin technology, the layer was readily evident as a structural keystone to serve for replicating the physical-virtual nexus. The second element functioning as the pivot point whereby all new reality is transferred into digital domain permitting conversion of information assets into functional satisfactions. As the point of convergence of our thorough investigate on the dynamic architecture digital twin structure, we distance ourselves ahead towards showing you how invaluable is Data Ingestion and Integration Layer: its powers featured as well as main in augmenting a data homogeneity; and eventually so required for immediate decision making.

It is true that the merger where real world signals become combined with digital insights occurs in a variety of applications, including models such as the digital twin and indicated by data ingestion and integration layer. This layer functions as a medium between the physical world and digital space, where accumulation of raw data turns to meaningful findings.

Here's a closer look at how this layer acts as the interface between real-world signals and digital insights:

Data Capture from Real-World Signals

This layer is responsible for capturing data from a diverse range of real-world signals, which can include sensor readings, telemetry data, machine outputs, environmental data, and more.

It could be accessing it from the outside is through physical sensors, IoT devices; databases and other sources represent data in different formats and protocols.

Data Preprocessing and Transformation

However, raw data mostly requires preprocessing and transformation to make it accurate, complete any compatible with the given digital representation. This involves activities such as cleaning, normalization and the filtration of data. Whereas real world signals may be noisy and therefore require some forms of signal processing to come up with pertinent information.

Data Integration
The layer unifies data from different sources into one consistent dataset. This involves matching of timestamps, normalization measurement units among others and eliminating any difference in the format used. Data integration provides a 'whole picture' of the physical asset or system as it combines data from different sensors and sources.

Real-Time Data Streaming
In real time critical insights, scenarios the layer could facilitate data streaming in near or true time. This allows for instant processing and analysis of incoming data hence quick decision making. So, in other words even the first sentence of 56. Data streams that arise out of human interactions, and which transport the reality or essence of real-life phenomena. This layer is the sentinel of physical environment because it encompasses temperatures readings from a manufacturing line or heart rate data from a patient. In so doing, the digital twin system gathers these signals to rise above common indication by providing a perfect representation of how current world situations work. Creation of a simulation, which creates an interactive link between the real world and its digital representation.

This layer is all about capturing data from many signals of the real-world information such as sensor-based thresholds, millimetric records, machine derived production and environment measures etc. It may also involve working with space sensors, IoT devices, databases and other formats and protocols that data extraction can be possible.

Data Preprocessing and Transformation
Data frequently have to be washed and transformed before it can provide a valid data that is comprehensive, truthful are the digital depiction. This involves activities like data harmonization, normalization, and profiling In the case of real signals present in the world, their respective individual observations are not only noisy but have to be premeditated by some form of signal processing techniques before any meaningful information can be recovered from them.

Data Integration
The data is aggregated from all the sources and built in a uniformed pattern to form a sole integral layer. Such aspects include stamp synchronizing, unification of units and resolution of issues regarding the format of data. The integrated data provided by the mechanism of data integration give a comprehensive view on physical asset or system in terms of historical elements comprising sense writing from sensors and sources that are plural.

Real-Time Data Streaming
In applications where the real time insights are critical, the layer may support in streaming Realtime data. This makes it possible for the reception and processing of information to take place almost instantly, promoting quick decision-making. There are sensors, IoT devices machinery resulting to the information that captures the identity of recital real-life endeavors. Whatever the environmental reading in terms of temperature from a manufacturing line or heart rates from a patient, this layer is

a sentinel both gathering and depending on the pulse of the physical world. The acausal bridge achieved by the digital twin system goes beyond simulation; with signals from reality captured, it has ushered into a new paradigm.

Seamless Data Convergence: The challenge of Heterogeneity In a world where all data sources are heterogeneous, it becomes necessary to fuse the disparateness and draw up a smooth plane for operation. The Data Ingestion and Integration Layer, therefore, shoulders the role of harmonization and ensures that heterogeneous data streams heed orderly convergence on one consolidated platform. It incorporates data transformation, cleansing and normalization that ensures all the editorial procedures speak a common language so it becomes easy to analyze and model. This level resolves discrepancies encountered resulting from integrated data of IoT devices or reconciles accurate insights consolidated from the different business units within an enterprise.

Operationalizing Data Governance: Quality and Compliance Data governance, which has not been emphasized enough in our discussion but is an overarching requirement enforced by the non-technical orchestration of the Data Ingestion and Integration layer. It is a development sequence that emphasizes the validation of data, handling of error, as well compliance standards. This layer also includes security features to prevent the loss of sensitive information during the ingestion process which reinforces reliability as well compliance.

3.2.2 *Analytics and Modeling Layer*

The sleeping twins are the Analytics and Modeling Layer because it is a central engine of the digital twin ecosystem that connects reality with its virtual forms in regard to tis real world which contains both physically and virtually. This essential level relies on the modern big data analytics and modeling that help to supply the ML-algorithms' performance ranges with value, insignificant perhaps because of high cost NPVs or when discovering complex unmeasurable relations for inferring needs. With the help of predictive analytics, machine learning algorithms, and simulation models it skips through just pointing out down what happened in the past and trends of future events but transforms into an interplay with digital twins to simulate real-life behavior. Such orchestration helps realize dynamic changes, what-if analysis, and prescriptive recommendations whereby organizations can keep adjusting performance of their operations by adopting predictive maintenance strategies in light of hybrids as such they will choose to make decisions anytime with the help from data. As represented by Analytics and Modeling Layer, continuous learning system is enabled in digital twins, maintaining that their representations manifest development of real-worlds [8–10].

Thus, digital twins are one of the vital elements for comprehensive application of digital twin technologies causing significant awareness among numerous industrial fields.

Insight Synthesis: At centre of the Analytics and Modeling Layer, alchemical synthesis Distillation is in manufacturing a factor out of phenomenon called its "data"; after this cast's perfect affinities mold. This project is transformative due to fact that a very wide spectrum of analysis techniques is used—from classical statistics to high-level machine learning procedures. This layer unravels significant shapes, trends as well as correlations from data streams it encounters on its way. When the sensor readings compare to information' that is stored in the digital warehouse, whether an outlier multivariable testing of normality or projection from purchase history data on customer preferences. This layer enables digital twin systems to turn raw information into actionable insights that steer intelligent actions.

Realistic Modeling: The layer that describe the framework of Virtual Reality is Analytics and Modeling Layer as it creates models which copy behavioral aspects characteristic nature of real-world objects. These models, usually conceptualized using physics and engineering concepts with mathematics being applied at their core, allow digital twins to perform simulations in various scenarios while offering predictions before an event takes place through the optimization of processes. For instance, through machine learning techniques such as production lines in the scope of smart manufacturing and can simulate this behavior to predict maintenance requirements or serve as an empirical estimate for variation's performance. This layer helps to reflect the same realities that occur in real life, so digital twins become not only objects of research but also allow for experimentation, optimization, and risk assessments.

Predictive Prowess: Tomorrow-Gazing From a technological standpoint, the Analytics and Modeling Layer is not one of the major successes achieved by Brute Force as its primary achievements include Peering into tomorrow's possibilities. This deck provides foresight by analyzing historical data and finding trends; being forecasting—able, this layer empowers digital models with the ability to predict future activity. In the field of supply chain management manages can make their own decision based on these predicted trends in order to achieve even inventory decision pre-empting shortage and surplus situations. This predictive power changes the traditions and strategies of the industries from adopting a reactive standpoint to pro-actively approach so that they can save on operational costs avoiding uncertainties.

Domain-Specific Optimization: Industry-specifics help the Analytics and Maintenance Layer to have Tailored Insights, which is beneficial for this layer. Modeling and analytics are carefully formulated tools to address the individual strengths and weaknesses of each field. In health sector, these models might be used to estimate readmission of a patient based on medical history and in the agriculture industry it could predict value harvest by checking climate information. Through creating a multilayer perspective, this platform targets upscaling its services based on the dynamics of the industry; hence becoming an innovation catalyst.

Cognitive Empowerment: Further into the trip of digital twin architecture, the Analytics and Modeling Layer is at the core while it enables insight from data that gives decisions from a model reality to predicts foresight. If we get to go inside the digital twin see what happens there, let us image that there is Analytics and Modeling Layer or smart part which turns data into information, makes abstract ideas real, and helps predict events. It is an instance of a listening translator who interprets the musical language into something you can hear and understand. This layer is good in understanding all the data our have and turn this data regarding into very useful insights that will be working well for various industries. Its core function of making the smart digital twin is highly essential since it enables that industries face and adapt to hidden shocks, capitalize on emerging opportunities, and make reasoned moves.

3.2.3 Localized Intelligence: Edging Insights in the Digital Twin Tapestry

Humanized and transforming the world of digital twin technology, we encounter an approach that creates a great sense of touch by taking localized intelligence like a brilliant touch to make the twin close copy live simulatable. This param rhythm takes us through a journey into understanding the significance of localised intelligence in the digital twin universe. We are also going to reveal its specifics, discuss how it facilitates snap judgement skills and investigate the way such industries as augmented reality are benefiting from making realities virtual [3, 11].

The Essence of Localized Intelligence: Local Intelligence is De-bifurcating Physical and Virtual Realms Intelligent localization centers on the fusion of environmental insight and real-time adaptability into the digital clone. Distinguished from centralized analytics that require processing to be done centrally, localized intelligence jumps onto the edge of network and allowing closer contact with nodes for real-time interaction. This intelligence is a model that replicates man's ability to think, enabling the virtual twin to be alive; it can respond variably when stimulated or adapt accordingly in a changing environment and make decisions altogether.

The Symphony of Real-Time Insights
This is when it may so do for in the era of intelligence at the grassroots, it seems smart only during intellectuals of time using information that will in most cases need short-term decisions to rely in the latest and correct piece of information. In sectors such as autonomation and industrial automation, active situational awareness through fresh sentinel input, machine sensors acting on the world and directing it into information, and virtual twins that give reliable feedback, predict impending failures, and timely alarms has become feasible, for coordinating rapid responses with localized intelligence the conductor. The combination of this world may allow a smooth sequence and accordingly most occurrences can occur whenever and however they ought.

Dynamic Adaptation and Personalization: Therefore, unquestionably, it is not an all-time ideal perfect-package solution required need for customization but a native solution that is depending on specific situations. Similarly, in the same way that an actor, intelligence, localized, answers what lift I can make and that in short decreases according to facts incorporated. For instance, in the health industry it supports focused patient surveillance to guarantee that they receive timely cure or prevention services according to those data available based upon future developments of research deserving further thoughts. Within a smart agricultural environment, local intelligence is related to digital twins acting as towards ensure the technique of irrigation and fertilization attains in order to build and overcome soil conditions or microclimates such as generated by masks.

The Industry Transformation Unveiled: All Industries May Indiscriminately Get to Feel the Surge of Samples with a Closer Look at Chic Particular Venus, Steel-meets-Applications. In manufacturing industry, it uses real-time application of quality assurance at the point of processing on production so that, there will be reduction in defects and wastefulness. Playing as data of real time toward perfect-pull consumption, localized intelligence is a peer to digital twins in adjustment of such parameters that represent power consummation based on the present electrical demand and supply. Whether local intelligence makes pending a shift towards refined functionality in smart cities as relevant to mobility today or pushes forth the power dynamics within and between these regions, it puts new flesh on what has remained till this day an undetected region—both ultimate types of power, physical and digital.

3.2.4 Paradigms of Architecture in Digital Twin Realization: Navigating the Pathways of Innovation

As such, the contributing paradigms of architecture within this incessantly developing context endowed with digital twin technology naturally appear as guiding constellations around which diverse routes to transformational solutions can be arranged. In this chapter, a discussion of the paradigms that permeate the architecture of digital twins will be undertaken, with an exploration of their propelling ideologies as well as how they address numerous challenges and serve as catalyst for industries looking to exploit the disruptive nature of this technology [1–4].

Monolithic Architecture: An Underlying Unity the Foundation underpinning digital twin architecture is the monad, a basic unit similar to how classically software systems were architected. Integration of all elements and functionality into a single monolithic structure is the key feature of this paradigm, which serves to strengthen ties between the components while making their management much less complex. Even though as an advantage, monadic architecture brings simplicity, scalability problems can be faced because digital twin systems grow to hold huge amounts of data and complicated interdependencies. Nevertheless, it is quite stable platform for the places organizations that embark on digital twin journey to launch themselves with an integration base solid foundation before venturing into different advanced paradigms [12].

Microservices Architecture: Microservices Mistranslate Managing Modularity and Agility The microservices paradigm dawns as an emblem of modularity and agility champing at the bit from monadic shackles. Within this paradigm, functionalities are splattered into various independent however lightly connected services that speak via APIs2. This decoupling allows for scalability to the effect that each component can independently be developed, deployed, and upgraded as needed. Microservices that are integral to the systems or organizations on their digital twin journey can help them alter and adjust components without having to recreate the entire system. This custom architecture increases adaptability, where data sources of visualization tools and analytics algorithms are easy to incorporate [13].

Event-Driven Architecture: Releasing the Response and Intentional Cohesion The event driven approach tweets as a cadence of intentional exegesis with reaction. In this architecture style, objects within the architectural components interact with each other using event that are either raised or consumed. But on occurrence of a significant event—like cutting across an interested threshold by some sensor fitting or in response to calibration calls from other parts and systems—occurring exchange, similarly, applies for enterprises making decisions based on reports created with the use of IT Tools – + architecture coordinates respective activities. This discipline has its niche in those scenarios where live analysis and dynamic adjustments officials accounts, semi-official type of role such as Head or Directors for External Affairs functions certainly need to be whistle distance if not near enough whilst clearly long-term planning considerations but high adaptation time. Manufacturing may identify with referent, as the equipment's break down will be sensed by policy and necessary maintenance procedures can help stop production line halting to make place for repair.

Edge Computing Architecture: However, under the popular titles of Handling Latency and Improve Realism Conduct The idea behind edge computing surfaces to supremacy is giving a promise that it will enable latency reduction brought across by localized intelligence. In this architecture, processing is done closer to the sources so that move-times towards a centralised cloud are bridged4. In use cases of digital twins that require on-time intelligence processing, such as driverless cars or remote-controlled surgery devices, edge computing ensures timely responses by allowing data to be processed almost in real time at the network's periphery. This paradigmatic ensures digital twins become more real through enabling them mirroring dynamics of the physical world as much close possible, thus further justifying timely authentic simulations and decisions.

Cloud-Native Architecture: The scope of solution and agility are also embraced by cloud-native architecture as a current emerging wave is that for cloud computing which is new architectural style. From this approach, these applications are deployed in a manner they can take advantage of the auto-scaling services offered by cloud during dynamic resource sharing fault tolerance as well [5]. With the increase in the data and user interactions within the digital twin systems, the cloud-native architectures support these facts better, and therefore, optimal performance rates are ensured although demand increases. Herein lies a paradigm whereby enterprises grow organically so that digital twin ecosystems grow in sync with the growth of the organization, thus allowing industries to grow infinitely while deriving economies of cost.

Through the descriptions of these architectures by one we realize that one feels the direction practitioners plot these separate domains of innovation. As different forms of art deliver a mixture of emotions, such paradigms would also release various capacities within a digital twin-sphere that range from improved sensitivity in event-driven architectures to microservice-based dexterity. This is why industries that capture the essence of their objectives as paradigm can offer another symphony of digital twin sovereignty that combines reality with virtuality through everything technique as the later element to deliver on operational agility.

Because of this, businesses that define their main goals as a paradigm can offer a new kind of digital twin autonomy that mixes reality and virtuality through the everything technique as the last part to deliver on operational agility.

3.2.5 Microservices Architecture: Pioneering Agility and Scalability in the Digital Twin Landscape

In the advancing landscape of digital twin technology, Microservices Architecture stands out as an outlier that shapes a new reality for coordinating the complex work of digital twins. This section deals with Microservices Architecture significance in digital twin, its architectural philosophy, it is represented as a complexity treatment approach and transformation ability when physical-to-digital processing becomes critical for a large number of industries that are seeking smooth interactions between these domains [13].

Architectural Independency: The core of Decoupling for Agility is exactly what makes Microservices Architecture the epitome of modularity and autonomy. In such a paradigm, capabilities of the digital twin system are statically segregated into individual services that encapsulate particular functionalities or processes. This is different from single structure designs that having something changed results in a cascade of changes, whereas micro services allow autonomous development and integration and such modules mature over time. Just like an orchestra comprises of diverse instruments, each with their tune that blend in harmony, so the Microservices Architecture becomes a set of serviceable aspects that complements each other to improve this unsegregated digital twin atmospheres.

Resolving Complexity: An important idea, microservices architecture, shows its power to handle the complexity of digital twin systems. In contrast, the diversity management in a large ecosystem like the sensors, information analysis and visualization is quite challenging in a single market.

Microservices force these much into development teams by wrapping facilities thus enable the team to handle separate channels of services that reduce the risk of complexity which lead to corruption. This architectural outline allows digital twin ecosystems to change gradually encouraging the practice of continual improvement and flexibility.

Scalability While empowering Growth, the quality of the Microservices Architecture is based on its performance that can grow in a remarkable way. When it comes to scaling in the world of digital twins, one of the top things that is unavoidable relates to an overwhelming increase in data volumes and intensified user interactions. This is resolved by the microservices, which ensure each component can be scaled separately and avoid over-provisioning resources while improving operational efficiencies3. The concept of microservices similarly allows the modulation of resource orchestration in response to changing demands, mitigating disruptions as a well-conducted adaptivity manager would balance each section's mutual volumes for harmony.

Agility Amplified It urges the spirit of innovation within digital twin systems so that embracing Innovation, Embracing Microservices Architecture is done. Development teams are able to try new things, take risks and create innovative features without affecting the stability of a whole system. This works perfectly in line with the agile culture, which supports quick iteration and flexible changes. For instance, within the health industry, microservices architecture would enable quick assimilation of new medical devices or diagnostic algorithms promising that the digital twin keeps pace with a fast-moving environment.

3.2.6 Modeling the Virtualization Layer

In the complex world of digital twin technology, we see how Virtualization and Modeling Layer comes in to play as an architect behind a parallel universe where physical realities are intertwined with its virtual counterparts. This section dives deep into what VML within DTM truly means. It examines the core ideas behind it and how it's used as a simulation and prediction tool across various industries. VML acts as a guiding light, helping these industries navigate complex challenges and gain valuable insights into the future [14].

Essence of Virtualization: VR in early stages is mostly the Virtualization and Modeling Layer, which kind of transits the reality into capturing the virtual surroundings. Italian [Models, most often, physical, mathematical and market-based arithmetic], help digital twins to function as their role-alikes.

Realistic Simulation Modelling any application from understanding a simple engine mechanism to the flow of the manufacturing processes using an automated guided vehicle (AGV) software, through traffic management on intersections, and endorsing air traffic control software one can visualize its operational processes as a lab test to validate the activity; assess an efficiency of exhibited patterns and refine the engineering prior to the deployment that in turn relieve many production concerns and lower the risk of failure.

For instance, in the aerospace applications this tier would model performance of an aircraft component given different scenarios hence correct on designs and effective use fuel.

Predictive Alchemy But as this data starts transforming into forecasting predictions an incredible phenomenon occurs in V-M Layer—the Out of Data layer gets attained. By combining the benefit of hindsight and prediction, digital twins promise to have insight into a look at future events in real-time by integrating historical information with predictive models. For example, in the health care sector these models can forecast early readmission of patients that may suitably be classified based on patient history and medical practice. Such predictive power helps organizations to anticipate trends, transcend or even forecast any upcoming problems and develop mitigation plans thereby changing the methodology from being reactive into proactive.

Dynamic Adaptation Permitting the functional transformations involved in evolutionary, choice of digitalization and Model Layer will surpass provide insights about change at work. Just as an observer at sea through live weather feeds manages to plot rough seas ahead of him, digital twins chart the exact waters using real-time data for live strategies. In the case of smart cities, this layer could optimize its function by using information on real-time and optimal flow scenarios at which it would intelligently control traffic lights to minimize road congestion. This contributes to the success of working life in respect with speedy action, resistance, and responsiveness.

The convergence of sciences and humanities does not wane when we visit the Virtualization Modeling Layer. It is in the intricacies of relations between data and models.

3.2.7 Case Study 1: Digital Twin of a Wind Turbine

It is in the heterogeneous realm that virtualization and Modeling Layer becomes as architect of simultaneous realities bridging between physical reality with its siblings, where digital twins come to full power carrying different operations. This para goes on a trip to highlight the profound significance retained by it within its relationship with entire nature, and in that process also determines its fundamental principles both herein as well as while representing or predicting future altering [14].

Essence of Virtualization: Virtualization and Modeling Layer is also the essence where Real world becomes yoga form, Virtualization. Simultaneously, this stratum is a painting to being as that famous artist who abridges soul of a particular scene on the simple fabric. Digital twin reflecting the behaving of real-world are empowered by such models, which in most cases come from physics, mathematics, and asset knowledge. Simulating any application—whether it is the engine mechanics; or flow dynamics of manufacturing processes, such as forklift operations in automated guided vehicles (AGV's), traffic movements at intersections and airports simulated by air-traffic control software devices, imaging machines running artificial intelligence auto-diagnosis algorithms predictively simulating what-if scenarios based on observed vital signs found to precede occurred malignant incidents—this layer

Realistic Simulation: Orchestrating Controlled Scenarios is developed by which the virtualization and modeling Layer regulates authorized simulations thereby driving industries into a realm of actuality. Similarly, to a flight simulator, which allows pilots rehearse aerial manoeuvres while lying on the ground; this stratum enables firms act out scenarios without entering into the real world. Digital twins can be used for modeling various scenarios based on historical data, from exogenous variables and certain predictions maybe helping with some models likely to predict future projections; digital twins that are able simulate different outcomes. For instance, in the aerospace applications this tier would model performance of an aircraft component given different scenarios hence correct on designs and effective use fuel.

Predictive Alchemy: But as this data starts transforming into forecasting predictions an incredible phenomenon occurs in V-M Layer—the Out of Data layer gets attained. By combining the benefit of hindsight and prediction, digital twins promise to have insight into a look at future events in real-time by integrating historical information with predictive models. For example, in the health care sector these models can forecast early readmission of patients that may suitably be classified based on patient history and medical practice. Such predictive power helps organizations to anticipate trends, transcend or even forecast any upcoming problems and develop mitigation plans thereby changing the methodology from being reactive into proactive.

Dynamic Adaptation: Leaving the functional transformations of evolutionary aside, choice of digitalization and Model Layer will exceed identify changes at work. Similarly, like the officer at sea with the help of live weather feeds finds himself able to plot rough seas in front of him, digital twins draw the exact waters with real-time data for live strategies. In the case of smart cities, the intelligence layer could optimize its operation by making use of data on real-time and optimisation flow scenarios at which time, it would intelligently manage the traffic lights so as to minimize road congestion. This also makes working life a dynamic success in terms of speed of action, resistance, and reaction.

The coming together of science and humanities does not fade away when we move to the Virtualization Modeling Layer. Just as a writer constructs fascinating worlds, which stimulate the imaginations of the readers, this unit creates an online reality that feed the imagination of consumers. The latter concept game power/displays and adaptation is the heart of this equation where struggling businesses twist through the complexity to drag on duck with unseen eyes orchestration like no other. It is in the detail of the relationship between data and models that In the case of a company producing renewable energy from a wind turbine farm with multiple units. Each of the turbines have sensors installed that take readings on wind speed, rotor availability and condition, temperature control operations for over heating systems like cables or generator regions combined with readings that advise respective power yields. The management would like to see these turbines perform the best and their controllability at optimum production level [15].

In this case, the Virtualization and Modeling Layer comes into use. By incorporating the concepts of physics, mathematics, and engineering into these systems; a digital replication of every wind turbine can be created. Models implemented in the digital twin simulate actual behavior of physical components of a turbine or its parts as well as interactions between these elements.

The process involves:

Physics: Using the basic principles of fluid dynamics, material science and mechanics to model interaction between wind flow with certain coupling conditions on turbine blades along stress strain path functions in other mechanical components as well number of heats dissipated from turbines structures.

Mathematics: This would involve using mathematical models that mimic how the blades interact with airflow, something called aerodynamics of a blade including torque generation by rotor and transformation from kinetic into electrical energy.

Domain Knowledge: Taking advantage of knowledge from that industry to adjust the models according to exclusive turbine designs, environmental situation s and maintenance activities.

After the digital twin has been created, real-time data from the sensors of actual turbines is fed into it continually. The digital twin makes use of such data to mimic the behaviour of turbine to make performance estimation as well fault identification.

- On further increase in the wind speed, digital twin can simulate changing of power output by turbine with respect to detect changes its value during operations preventing operator from efficient use.
- More so, if the temperature of a particular component goes higher than some level determined upfront, its digital twin may show what happens to this specific part in simulated time and suggest whether anything should be done for maintenance.
- The digital twin then compares the information obtained from different turbines, this helps it to notice any inconsistencies and point out possible malfunction or inefficient performance of a turbine.

The above example of the case shows how VML makes it relevance hail between phenomenon wind turbines and virtual presence. The digital twin makes the business optimize energy generation, project instances of preventive and reactive maintenance for operational efficiency using real-life behavior based on physics mathematics or domain knowledge.

Modeling Aerodynamics: Wind turbines owe much to the science of fluid dynamics. Modeling the flow of air across turbine blades is based on the Navier-Stokes equation which is a mathematical tool for representing movement of fluid. Solving these equations in Computational Fluid Dynamics (CFD) software generates numbers that can be used to predict such effects as lift and drag forces acting on the blades. This information is very significant in the comprehension of the rotation and energy production influence of wind speed.

Mechanical Stress Analysis: Turbine components experience stresses and strains due to rotational safe forces or wind loads, which are modelled using solid mechanics principles. Finite Element Analysis (FEA) method is a mathematical method often applied to consequences of mechanical behavior. FEA algorithms are created to divide turbine structure into small elements and analyze stress distributions, hence, these systems identify critical areas that result in fatigue or breakdown.

3.3 Simulation of Power Generation and Rotor Dynamics

Power Generation Simulation: The upper value of the energy, which can be determined by an upper bound called Betz and guided by technical parameters of ideal rotation conversion of kinetic power, is available. This threshold is used by the algorithms as a simulation of the wind speed that is available for power generation to inform operators of how different winds affect their energy production levels.

Rotor Dynamics Simulation: When rotors turn the turbine, they have a set of dynamics that can really be related to mathematical functions. To accomplish such tasks, motion equation-based algorithms are used, for example, finite difference method make the possible solution of how speed changes with time at a rotor taking into account inertia and torque. It is useful when establishing the turbine performance in the range of the wind speeds and loads.

Component-Specific Models: Domain specialists instead will provide specific, detailed and justifiable attributes or characteristics by taking the viewpoints of how a material is likely to behave between them offering what can be used during design considerations as well dimensionalized information based on historical data they have accumulated over time. These incentive material properties used for turbine blades, gearbox and bearings components. These properties are added to the models being used by digital twins for simulation, making it relatively perfect as how things happen in real life.

Calibration and Validation: Models of the digital twin would be roughly course validated, and calibrated may be by domain knowledge. To check this, the historical data of the turbine in its mechanical form such as different power output at different winds is used to re-valuate these algorithms and that those predictions generated by the digital model indeed match actual performance. Production of the digital twinning process is done iteratively with the involvement of domain experts who ensure that results from a specific intervention correspond to realities.

Algorithmic Workflow: Combining Physics, Mathematics and Domain Knowledge

Aerodynamic Simulation: Use CFD algorithms to model the airflow over turbine blades. Numerically solve the Navier-Stokes equations for lift and drag forces calculations.

Mechanical Stress Analysis: The stresses of the mechanical work are modeled using FEA algorithms in turbine components. Segment the structure into parts, and calculate stress distributions from mechanical properties.

Power Output Simulation: Simulations of power generation with the help of wind speed data are performed through mathematical equations that include such formulas as knowns at limit in Betz.

Rotor Dynamics Simulation: Develop algorithms to address rotor dynamic equations of motion. Consider inertia, torque, and wind load to make predictions in changes of rotor speed.

Incorporate Domain Knowledge: Incorporate the material properties obtained from domain experts in performing simulations. This guarantees that when creating the model, component behavior is adequately imitated.

Calibration and Validation: Calibrate algorithms and validate forecasts with historical data. Adjust the parameters of digital twin to improve matching actual turbine performance.

Continuous Improvement: Include real-time information coming from turbine itself into your digital twin. Modify algorithms with fresh observations as a way to increase accuracy over time.

Thus, the Virtualization and Modeling Layer makes a digital twin that accurately reflects how an actual wind turbine works by capturing both mathematical and algorithmic approaches within the wisdom of subject matter experts. This enables personnel to unlock knowledge about power generation, predict timely maintenance needs and better optimize operational plans that are incredibly accurate [24].

3.3.1 Case Study 2: Create a Digital Twin of a Wind Tunnel

Modeling Airflow A wind tunnel is a facility that allows for the creation of conditions relatable to those encountered by an object during flight in terms od aerodynamics. The flow of air within the tunnel can be analyzed with much resemblance to fluid dynamics. Navier-Stokes equations governing fluid flow may be numerically solved to model airflow in the wind tunnel. As such, this digital twin even enables forecasting how the air environment will behave around an object model installed in a tunnel [16, 17].

Pressure Distribution Simulation: Regarding aerodynamic characteristics of the object, its pressure distribution on the surface in a wind tunnel is very important. Based on the flow velocity, air density and shape of an object mathematical algorithms can compute how pressure distributed. This data allows us to predict the lift and drag forces that are generated on a model.

Drag Calculation: One can devise a quantification of the drag force acting on this object by means of a characteristic coefficient Cd and dynamic pressure q, as follows

$$\text{drag} = 12\ \text{Fdrag} = 21\text{CdAq}$$

Where:

- A is the object reference area.

Object-Specific Models: Objects have different aerodynamic shapes. Also, as the information for this field of study domain expert fills in geometry and surface properties of object under examination along with possible areas for turbulence. This fact is implemented into the simulations of Digital Twin (used in order to simulate some processes) through which it replicates behaviour common for physical wind tunnel.

Calibration and Validation: Similarly, in analogy to the wind turbine example, historical data and experimental measurements from the physical earned can be utilized during calibration and validation. The parameters of the digital twin can now be adjusted to get a better result by referring to observations in reality.

Algorithmic Workflow: Fusion of Physics, Math and Domain

(i) *Airflow Simulation:* Use numerical techniques to solve the Navier-Stokes equations and model wind flow in the tunnel. Anticipate the air flow in relation to this model.
(ii) *Pressure Distribution Simulation:* Utilize mathematical algorithms to determine the pressure distribution on the model surface. This data provides information to predict lift and drag forces.
(iii) *Drag Force Calculation:* Knowing the drag coefficient and dynamic pressure, one can find a linear force acting on that model in the wind tunnel.
(iv) *Incorporate Domain Knowledge:* Incorporate domain-specific knowledge of the geometry and properties of object A into the simulations in order to improve verisimilitude.
(v) *Calibration and Validation:* Calibrate and validate the digital twin predictions using experimental data derived from physical wind tunnel tests. Fine tune some of the parameters to match the output from DT with real world observations.

Mathematics, physics, and domain knowledge are used in the process of designing a digital twin of any wind tunnel. The digital twin creates a simulated model of the object and using certain computational tools, analysis for aire flow, pressure distribution prints are created inside cyber space that help determine what will happen to different objects in the wind tunnel. This provides researchers and engineers with opportunities for design optimization, performance analysis, decision making based on established facts without testing in the real physical world.

Algorithmic Workflow: Digital Twin for Wind Tunnel
Input:

- Tunnel geometry and dimensions
- Flow velocity V
- Air density ρ

Process:

- Use computational fluid dynamics (CFD) algorithms in order to numerically solver the Navier-Stokes equations and, thus, simulate both flow field inside a wind tunnel.

Output:

- The measurable vectors on the tunnel's instantaneous direction(t) within(x, y, z), which is denoted as velocity field$\vec{v}$(x,y,z,t).

ii. Pressure Distribution Simulation:
Input:

- Tunnel wall properties

Process:

- Calculate the pressure distribution which is developed on the outer surface of model in wind tunnel using mathematical equations.

Output:

- Surface pressure field (x, y, z,t) of the model.

iii. Drag Force Calculation:
Input:

- P(x, y,z,t)
- Concerning Model, where A is the comparator region
- Dynamic pressure = 12 2q = 21ρV2

Process:

- Find the drag coefficient Cd from pressure changes products and known relations but not directly.
- Determine the drag that develops on the model Fdrag = 12F_F21CdAq.

Output:

- The force of realistic drag acting on the model.

iv. Incorporate Domain Knowledge:
Input:

- The object specific properties are domain expert provided (geometry, surface variant characteristics like roughness or drag coefficient oscillations and regions inducing the turbulence)

Process:

- Introduce region directions in the simulation algorithm with intention to enhance on their accuracy and realism.

Output:

- Functions of the specific characteristic on or around an object that is solely based on outcomes presented in a simulation environment.

v. Calibration and Validation:
Input:

- The wind tunnel data that were measured identified in full—scale operation (i.e.though pressure measurements, as well as by measuring drag forces on the structure of rotor blades) speed distribution etc.

Process:

- Adjust the parameters of simulations for agreement between predictions by digital twins and conclusions based on experimental observations made in reality.
- Fine-tuning the algorithms must be conducted to observe more consistency between results generated from twin and wind tunnel data.

Output:

- Self consistent case study simulation models—which are calibrated and validated to capture actual wind tunnel dynamics.

As a result of this formalized algorithmic workflow, principles borrowed from physics and mathematics in addition to domain knowledge are assimilated into the digital twin as it simulates flow fields, predicts pressure distribution calculates aerodynamic forces finally producing information regarding object behavior within an environment built on wind tunnels.

3.4 User Interface and Interaction Layer

The User Interface and Interaction Layer is a component of the system architecture, which constitutes interaction channels between users and their digital twin virtual environment. Interfaces, the interfaces and people's feelings of visualizations are also rather comfortable to be used making possible human receive an immediately understanding at them in order for him to make decisions based on their results [18–20].

3.4.1 *User Interface and Interaction Layer: Enabling Seamless Digital Twin Participation*

In a more granular scale of the digital twin architecture, User Interface and Interaction Layer embeds invisibly deep between these computational layers as an insensible cornerstone which fusion human cognition with combinatorial acumen. This chapter penetrates the hall this level, detailing how intuitive visualizations operate and receptive dashboards frame works as well as immersive interfaces where users can journey down routes through the digital twin realm liberating insights and embarking on data-driven excursions.

Visualizing Complexity: A Symphony of Data and Design
Visualization lies on the side of User Interface and Interaction Layer or can be explained as all techniques about how to present data in such a form that would be comprehensible for human being. An algorithm stands at the center of this choreography and is utilized to translate highly dense data sets commonly produced in real-time by sensors or via simulations, into palatable visualizations. This is achieved through adoption of certain data processing techniques such as the likes coagulation, interpretation, and scaling to numbers of multivariate pictures (including graphs or charts; heat presses maps 3D models) where anomalies trends well relationships are obvious.

Responsive Dashboards: A Portal to Insightful Discovery
By means of this User Interface and Interaction Layer, workbenches which become the doors to information treasures stored in the digital twin property can be shared with organization. These dashboards gather all the visualizations used and make them active compositions to offer users a way of customizing their investigation. Nevertheless, a web framework is supported by front-end libraries that work hand-in-hand with data-binding algorithms handling the dynamic interaction. The visualization tool enables the user to apply filters on datasets, alter parameters and switch between models as they simultaneously observe how changes in inputs echo through matching hollow results of twin. These interactive dashboards are therefore turned into cyberspace labs where hypothesis is verified, scenarios get full evaluation and conscious decisions made.

Immersive Interfaces: Bridging the Digital-Physical Gap
Thus, immersion interfaces are coped into the scope use of The User Interface and Interaction Layer thus attracting to its limits never-been paradigms of interaction. Digital twin acts as a platform for immersive experiences when technologies like Virtual Reality (VR) or Augmented Reality are involved. Even when dealing with tangible DWs, the process has turned into an entirely new level thanks to virtual reality headsets which enable direct work on them that are placed elsewhere transcending Two-dimensional Display Units (du).

For example, in the manufacturing processes engineers would be the right category of people to watch such complex parts of a machine, and they can probably imagine the digital information as the physical things while mentally interpreting real data streams in terms of the virtual overlays. Since, such a fusion of virtual and real entails to merge the separability between conceivability as well as reality and provides in an interesting dialogue which occurs between participant's and their virtual avatars.

User-Centred Design Versus Technical Accuracy

User Interface and Interaction Layer reflects the chord in the world of user centered design principles which is almost technical alarmingly. By developing critical norms for structure, colour shading use on environments and composition with the present-day advancements, for example, JavaScript designs, information handling libraries, and sign acknowledgment calculations. Every pixel reveals the cognitive ergonomics—intuitive but clever interactions; hygienic standard now and information available to gurus but also novices browsing digital twin landscapes.

Synthesizing User-Centred Design and Technical Precision

The algorithmic coherence begins with complex data sets being fed into this virtual world. The underlying data created by physical assets is fed in real time, and the consequences are determined using high-fidelity approximations.

The algorithm that precisely converts these complicated and typically real-time flowing datasets from sensors or simulations to compelling visual expressions for perception is at the heart of the User Interface and Interaction Layer. This method is sophisticated, requiring data visualisation and adaptation of raw algorithm outputs [21, 22].

This initial chaos news glut soon becomes a raw material for the choreography. Similar to virtuoso conductors, the algorithms process this information through strange formats correction harmonizing and anomalies timestamping. The outcome is a coordinated set of data points ready for the next step.

Dimensionality Reduction and Aggregation: Crafting Nuanced Narratives

With the development of dance, intricate algorithms for dimensionality reduction and aggregation make an appearance. These algorithms condense the many dimensions of data into non-monotonous tunes well, methods such as the Principal Component Analysis (PCA) can simply and concisely describe leading trends that allow for a reduction of complexity without losing important data. An aggregation algorithm, on the other hand, simplifies this symphony into brief melodies; it is a summary of variations and both spatially or temporally related.

3.4.2 Techniques for Ingestion of Diverse Datasets

The ingestion of diverse datasets is a crucial step in creating a digital twin's User Interface and Interaction Layer. Integrating a wide range of data sources, formats, and frequencies requires effective techniques to ensure data consistency, accuracy, and accessibility. Followings are several techniques that are useful for ingesting diverse datasets [23, 24]:

A. *Data Integration Platforms:* Data integration platforms provide connectors, mapping tools and transformation capabilities in order to join this data from disparate sources. Such platforms offer a ready-to-use integration component for popular data sources such as databases, APIs and cloud services. They include such systems as Apache NiFi, Talend and Microsoft Azure Data Factory.
B. *Extract, Transform, Load (ETL) Pipelines:* ETL pipelines include the process of data extraction from source systems, their transformation in order for them to be consistent with a target schema as well loading into the destination database. ETL tools help cleaning data, transforming it and enriching. Some popular ETL frameworks include Apache Spark, Talend and Apache Beam.
C. *Data Virtualization:* Data virtualization tool enables end-users to have a single integrated view of one or both physical and logical data from heterogeneous sources without any copying it physically. These tools create a virtual tier that enables the interrogation and retrieval of data from different sources as if they were contained within one dataset.
D. *Application Programming Interfaces (APIs):* A wide range of data sources provide an API that can be used to make program-generated requests for the provision of structured information. Some other APIs that help retrieve data from different sources, including web services, third-party applications and cloud platforms include RESTful APIs Graph QL SOAP-services.
E. *Message Brokers and Publish-Subscribe Systems:* For example, in Apache Kafka or RabbitMQ message brokers that can be used for sending, receiving and processing of the data may not necessarily align with receiving and registering creating a consistent update rate. This is especially convenient for the consumption of live data feeds from sensors, IoT devices, and other origins.
F. *Web Scraping and Crawling:* Web scraping is used for consuming data from websites, and different types of web crawling techniques are applied. These methods comprise automated structuring of information from html pages, xml files or any other structured content. Beautiful Soup and Scrapy are libraries that assist in webs scraping.
G. *File Formats and Data Serialization:* The data in the various alternative file formats (CSV, JSON, XML Parquet primitives), requires parsers and serializers, to allow their structure orchestration utilizing a standard form for processing machination.
H. *Data Preprocessing and Cleansing:* Data often need to be pre-processed and cleaned for validity before they can pass through the entire gross motor system. This can include managing null values, eliminating duplicate entries and standardizing data structures.
I. *Schema Mapping and Transformation:* Schemas of data can be different in case they are obtained from different sources. Referred to as schema mapping and transformation tools, these schemas can be transformed into a common format so that they integrate without any problem.

J. *Custom Data Wrangling Scripts:* For data sources that are novel or have non-standard formats, it is necessary to write scripts or programs for wrangling and transforming the said data. It is often implemented with Python, Rand other scripting languages.
K. *Data Quality and Validation:* The integration of controls and validation techniques in the load helps to ensure that verifiable data infiltrates into the digital twin environment.

The selection of technique is also based on data type, the update frequency and source system integration complexity.

3.4.3 Communication and Integration Layer

In the complex work of digital twin architecture, Communication and Integration Layer is revealed as a key mediator which connects different insights on data that can be provided at every level such as from systems. This section begins a discussion into the complexities of this middle layer, revealing its inner workings that support data communication and interoperability between different components throughout digital twin hubs [25].

The APIs, MQTT (Message Queuing Telemetry Transport), and RESTful endpoints are the languages of translation that allow data conversations to happen. These protocols support safe transmission of live data, which makes sure that the behavior of the components interacts and works together across a digital twin environment.

The standards like, OPC UA (Open Platform Communications Unified Architecture) and FIWARE provide the same dialect that different systems starting from diverse backgrounds can speak together and work.

The Communication and Integration Layer get synchronized with the event management time signature. In this model, actions create ripples in the downstream events that spread across the digital twin ecosystem to provoke reactions in other elements. The brokers such as Apache Kafka and AWS Event Bridge play the role of the pipes down which these events flow, ensuring that an action in one part of the digital twin resonates across its whole landscape. Such synchronicity makes the actions and reactions twins that are alive.

3.5 Conclusion

This chapter explained the basic architecture and design decisions that are fundamental in the digital twin technology context. In this comparison, between the blueprint in building design and architectural considerations of digital twins, we have shown how these decisions define the innovation pathway that is followed by including effectiveness, scalability, and adaptability. The attention is given to the building of a robust data architecture because data is coming from various sources.

In other words, that an ideally shaped data architecture is digital's twin backbone system supported what was discussed in this chapter regarding perfect connectivity and insights beyond domains. While discussing DN-based implementation, shift to understanding and stating concerns regarding these solutions are identified although not elaborately in relation to scaling such implement Addressing this dynamicity nature of digital twin implementations imply correct architectural decisions have direct dynamics impacts on scalability-performance. Responding to the increasing complexity and magnitude of operations, scalability considerations such as distributed computing and responsive user interfaces are crucial for ensuring responsiveness as discovered. Essentially, this chapter acts like a primer that lays the groundwork for understanding decisions that players must make when crafting digital twin architectures in an interconnected manner. Enabling such organizations to engage with both the virtual and physical realms necessitates holistic evaluations of architectural and design aspects that would lay a strong foundation to support innovation, integration as well the realization of organizational goals in an ever-changing digital twin arena.

References

1. Singh M, Srivastava R, Fuenmayor E, Kuts V, Qiao Y, Murray N, Devine D (2022) Applications of digital twin across industries: a review. Appl Sci 12(11):5727
2. Wang K, Wang Y, Li Y, Fan X, Xiao S, Hu L (2022) A review of the technology standards for enabling digital twin. Digital Twin 2:4
3. Lv Z, Xie S (2022) Artificial intelligence in the digital twins: state of the art, challenges, and future research topics. Digital Twin 1:12
4. Digital Twin Architecture & Standards – Visartech Blog, Visartech (n.d.) [Online]. Available: https://www.visartech.com/blog/digital-twin-architecture-guide/
5. Anon (2022) Digital Twin Architecture & Standards – Visartech Blog. Visartech, Available at: https://www.visartech.com/blog/digital-twin-architecture-guide/
6. van Dinter R, Tekinerdogan B, Catal C (2023) Reference architecture for digital twin-based predictive maintenance systems. Comput Ind Eng 177:109099
7. Tekinerdogan B, Verdouw C (2020) Systems architecture design pattern catalog for developing digital twins. Sensors 20(18):5103
8. Tekinerdogan B (2023) On the notion of digital twins: a modeling perspective. Systems 11(1):15. https://doi.org/10.3390/systems11010015
9. Kukushkin K, Ryabov Y, Borovkov A (2022) Digital twins: a systematic literature review based on data analysis and topic modeling. Data 7(12). https://doi.org/10.3390/data7120173
10. Ogunsakin R, Mehandjiev N, Marin CA (2023) Towards adaptive digital twins architecture. Comput Ind 149:103920
11. Shu M, Sun W, Zhang J, Duan X, Ai M (2022) Digital-twin-enabled 6G network autonomy and generative intelligence: architecture, technologies and applications. Digital Twin 2:16
12. Ferko E, Bucaioni A, Behnam M (2022) Architecting digital twins. IEEE Access 10:50335–50350. https://doi.org/10.1109/access.2022.3172964
13. Steindl G, Kastner W (2021) Semantic microservice framework for digital twins. Appl Sci 11(12):5633. https://doi.org/10.3390/app11125633

14. Rodrigo S, Mario DR, Moreno JI, Àlvarez-Campana M, López DR (2023) Digital twins for 5G networks: a modeling and deployment methodology. IEEE Access 11:38112–38126. https://doi.org/10.1109/access.2023.3267548
15. Fang F, Guishan Y, Yang H, Xutao W, Jizhen L (2022) Digital twin system of a wind turbine. Scientia Sinica (Technologica) 52(10):1582–1594
16. Renganathan SA, Harada K, Mavris DN (2020) Aerodynamic data fusion toward the digital twin paradigm. AIAA J 58(9):3902–3918. https://doi.org/10.2514/1.j059203
17. Mishra R, Guilmineau E, Neunaber I, Braud C (2024) Developing a digital twin framework for wind tunnel testing: validation of turbulent inflow and airfoil load applications. Wind Energy Sci 9(1):235–252
18. Newrzella SR, Franklin DW, Haider S (2022) Three-dimension digital twin reference architecture model for functionality, dependability, and life cycle development across industries. IEEE Access 10:95390–95410. https://doi.org/10.1109/access.2022.3202941
19. Boyes H, Watson T (2022) Digital twins: an analysis framework and open issues. Comput Ind 143:103763. https://doi.org/10.1016/j.compind.2022.103763
20. Aheleroff S, Xun X, Zhong RY, Yuqian L (2021) Digital twin as a service (DTaaS) in industry 4.0: an architecture reference model. Adv Eng Inform 47:101225. https://doi.org/10.1016/j.aei.2020.101225
21. Jia W, Wang W, Zhang Z (2022) From simple digital twin to complex digital twin part I: a novel modeling method for multi-scale and multi-scenario digital twin. Adv Eng Inform 53:101706
22. Jiang L, Su S, Pei X, Chu C, Yuan Y, Wang K (2022) Research on product-part level digital twin modeling method integrable to digital thread framework. SSRN Electron J
23. Jones D (2021) Artificial cognitive systems: the next generation of the digital twin. An opinion. Digital Twin 1:3
24. Ainsworth C (2023) Building your digital twin. New Sci 257(3432):34
25. Jadhav SG, Sarnikar S (2023) Digital twin of a digital world: process, data, and experience perspectives. IT Professional 25(3):68–73

Chapter 4
Digital Twin: Applications

4.1 Introduction

Digital twin technology is effectively changing almost all industries worldwide into a powerful tool for operations, efficiency, and decision-making. This chapter discusses the different uses of digital twins along with industry-specific case studies. Digital twin technology was adopted early in the manufacturing industry to improve the production process, minimize downtime, and improve product quality. For instance, Siemens has digital twins for its gas turbines and has developed intricate virtual models.

Such models allow online monitoring, predictive maintenance scheduling, resulting in 20% decrease in unplanned downtime and 10% increase in turbine availability. The energy industry is currently relying on digital twins to optimize energy consumption, cut emissions, and boost grid resilience. A digital twin platform for power grids is developed by Schneider Electric, which is a global leader in energy management and automation, to help in providing real-time monitoring and predictive analytics to detect and prevent potential failure situation before outages take place. Healthcare is using digital twins to create personalized medicine and increase patient treatment. In the realm of health care technology, GE Healthcare has created digital twin technology for medical tools such as ventilators and anesthesia machines. Such digital twins enable healthcare providers with monitoring device performance in real time, detecting of potential malfunctions, and settings tuning to each patient's needs.

In the transport business, safety measures, are also changing due to the use of digital twins providing alternatives to real applications, thus allowing for reduced road traffic congestion as well as optimized logistics. The Siemens has collaborated with Munich to develop a digital twin of the traffic system of the city this allows the city planners to simulate various consequences of the traffic. The above-mentioned digital twins optimize LED timing and execute the action of congestion mitigation. Digital Twin: case studies and reviews provides an insightful investigation of the

© The Author(s), under exclusive license to Springer Nature Switzerland AG 2025

S. Gupta et al., *Digital Twins*, https://doi.org/10.1007/978-3-031-76564-3_4

various uses of digital twin. For instance, Ayerbe et al. [1] analyze the digitalization of lithium-ion battery manufacturing operations, offering a three-layer digital twin framework for vast-scale battery production. This particular case study shows how digital twins can help increase the efficiency of manufacturing systems and improve the standard of quality control.

Ozturk [2] presents a discussion on the establishment of digital twins in construct and facility management which helps to manage buildings and infrastructure. A rendering of the present requirements is offered by the study by Wagg [3]. Furthermore, it is indicated, that as Löcklin et al. outlines in their article, digital twins can be a tool for verification and validation of industrial automation systems (2020). It shows to what degree digital twins can enhance the quality and reliability of the process. The Diagnosis and Prognosis topic discussed by Ibrahim [4] touches on the role of digital twins associated with LED, one of the vital components in terms of reliability and performance. Schrotter and Hürzeler [5] highlighted the digital twin of Zurich for city planning and showed its use in the management and development of urban areas.

Digital twin applications can offer several benefits like instantaneous monitoring, forecasting, enhancement of performance, and unique proposal. However, it is associated with numerous challenges that include the intricacies of data gathering and integration, escalated model complication and protection, and secrecy concerns over the big data implicated in the situation in question. In the field of industry, the technology used in digital twins is rapidly changing operations, increasing efficiency and supporting informed decision-making processes across the world. That is, this chapter implies a more applied nature of the digital twin with the use of the cases, samples of real digital twin implementation which is industry-specific and constitutes a major influence of the digital twin.

With those in the sector of global decision makers they only work with digital thinking of the type of products they would like to get and as such, the idea of digital twins has become crucial. It is besides manifestation that even if the benefits are laid down their implementation is faced with various challenges that needs to be addressed which include the data complexities, integrations to the current systems and safe security features that may not be comprisable. Nevertheless, it is obvious that with an increasing number of publications and research projects related to this field of study more attention is going to be paid to the topic of digital twins, which requires a thorough review of possible application areas and benefits that digital twins can offer.

4.2 Application of Digital Twin and Industry

Increasing efficiency is one of the targets that makes the use of digital twins very appealing in the manufacturing industry. These 3D holographic simulations of industrial processes speed up the process of evaluation and analysing data related with operation optimization. In terms of the assembly lines, the digital twins allow enterprises to track such crucial indicators as temperature, pressure, and speed, along with environmental factors. This control is useful for uncovering anomalies in the production process, which may indicate the presence of some problems or defects.

To clarify, in the case with process manufacturing, such twins control batch processes aimed at ensuring the best quality possible, predicting equipment failures using the principle of machine learning, and guaranteeing the safety and regulation standards apply in nearly real-time mode. In the power sector, digital twins are, especially used in forecasting consumer energy demand, using the artificial network models, to strengthen grid transmission and management, to support solar array management, and to support futuristic repair of wind farms through an insight they provide about the maintenance requirement. Real-time analytics are used to evaluate the integrity of the offshore structures in oil and gas, refresh the reservoir models using the exploration data, as well as monitor equipment such as pumps and compressors for better performance and maintenance.

Digital twins are of relevance to the mining and metals industry by enhancing recovery yields in mineral processing, waste monitoring, and offer casting guidance in metallurgical processes.

In the automotive sector, they find their use in designs of vehicles to vehicle end visibility, real-time telemetry interpreting to maintenance services and encourage the development of new business models in autonomous vehicle services.

Where a life sciences and medical applications of a digital twin will help in reducing stock and logistic risks, minimize downtime when dealing with high-performance systems and give holistic views of a patient though privacy and security will have to be considered. In the case of infrastructure, it avails off-site and on-site pre-fabrication, which assists with emergency coordination, boosts response times and records foot traffic in shops that are retailing. In the aerospace domain, digital twins support material and supply chain management, the increase in aircraft components' durability, and the revenue augmentation through airport operations optimization.

Specifically in defense, they provide with a well increased reliability with complicated gears that are used in the military to help in strategies by offering real time data and maintains a unified and a very lively series of data that ends up being needed by the mission critical schemes. Digital Twin is a digital replica of the physical entity, system or process and has very broad applicability across the industry spectrum.

However, with the onset of the Internet of Things era and pervasive data, and empowered by a revolutionary technology, the line between the physical and digital world is blurring.

(i) Manufacturing: Picture an assembly line before you where machines inform you of their wellbeing, cause of loss due to machine wastages is predicted, and manufacturing processes re-organized instantly. This is a very achievable fact in the manufacturing process through the use of Digital Twins. Data is fed into virtual twins through sensors, which contain those within equipment, performance, degradation, variables and so on and so forth besides the AI algorithms both process the same contents which have been built into it and forestall it on what it believes will go wrong before occurrence of any incidents. Therefore, reduced downtime, increased overall productivity and improved the quality of the product.

(ii) Aerospace & Aviation: Aircraft utilizing such advanced Digital Twins are found in airplanes that fly beyond the regions of the cloud high up in the sky. These twins manage nearly everything of the engine, encompassing control variables, fuel variables, environment's level of interest variables, including the flight mechanics, too. Hypothetical productions and forecasting of potential issues in different states use well hence helping in the prevention of various dangers to those individuals, most importantly the pilots, the best flight pathways, and guaranteeing security to the tourists and relief personnel.

(iii) Healthcare: This simulation of virtual patient is advancing the state of the art in the healthcare owe of the Digital Twins. Based on the part of a patient's archive and the recent state of human body conditions, such complicated models enable doctors to individualize actions, provide prognosis on likely complications and even conduct an operation virtually. For example, consider the possibility of performing a newly created treatment for cancer by a Digital Twin of a patient to determine its efficiency and negative consequences for a long time before applying it to practice.

(iv) Smart Cities: Consider a city that inhales this data treatment. The traffic lights change because of the pattern of movement at that instance, waste collection only takes place once the waste bin is full, and the energy grid forecasts and controls demand variation. Thus, the key concept of the Digital Twin City is to create digital twins of the relevant elements of infrastructures that will allow for the control of the coupled infrastructures and suit{s} of intervention measures as countermeasures against the above-mentioned urban challenges.

(v) Agriculture: What we observe is the change of the nature of such an industry as agriculture—Smart Farming or the Smart Revolution, Data-driven Digital Twins turn into a disordered work organization farmyard. Data obtained through the sensing devices such as about the soil, crop or feed provided to the animals is input to digital structures emanating information concerning the status of the plant, the plant's soil condition, the soil moisture levels and state of the animal. It will then help the farmers in getting to know the best irrigation methods, amounts of fertilizers to be use and the control measure of your livestock heard hence increase in crop holding potential as well as impact to the environment.

(vi) Supply Chain & Logistics: Picture a supply chain that is so open that you will be able to track a mango out of its growth cycle, all the way to the time it lands on your morning table. In the area of logistics, digital twins support the management of transportation of cargoes, decision routing, prediction of delaying, or signaling integrity breaches. This essentially results to delivery at a faster speed, cost savings and again customers fulfilling.

(vii) Energy & Utilities: Power grinds. Prepared to respond by shaking due to music data. Through provision of information regarding the workability efficiency, provide forecasting on the future inefficiencies, and find out how to optimize energy production and distribution; Digital Twins of energy cross section wind turbines to power subs deliver real-time self-produced intelligence. This results in the development of a stronger current network that is also less wasteful and renewable energy friendly.

(viii) Beyond the Horizon: Nevertheless, the application range of Digital Twins is far wider than was overviewed. The discussed ones are part of the cultural heritage site preservation, and it ends by the efforts of making cities more resilient to natural disasters, it keeps on increasing. However, because the ability of data collection and AI algorithms, in general, will only continue to improve, more disturbing use cases should emerge soon.

The compatibility of Digital Twin to unite the real world and that can achieve optimization, prediction and innovation than was ever imagined opened new avenues. We are not only witnessing with this game-changing technology, a whole new horizon for everything from efficiency, sustainability and right to the improvement in quality of life in just about every industry one can consider.

4.2.1 Manufacturing: Revolutionizing the Factory Floor

Digital twins in manufacturing can be used to streamline production lines, manage the condition of the equipment, and predict maintenance. It helps manufacturers to virtually simulate production processes and test them before implementing them in the real world This improves efficiency and reduces downtime.

Digital twin is the digital replica of a physical asset or a system which is in the form of a dynamic visual presentation and is updated based on real time to represent its workings and characteristics. Here the stage is set for an ideal feedback loop, wherein the physical, as well as virtual, intertwine, enabling manufacturers to attain hitherto unimaginable levels of optimization, forecast-driven, and innovation.

The effect of digital twins upon the manufacturing process is complex and obviously positive. Consider the following benefits:

Predictive Maintenance: The sensors integrated into the machines constantly transmit data that relates to the parameters of the equipment's performance, the degree of the wear and tear of the equipment, and the work environment to the digital double. The AI algorithm process such information and detect possible failures and the need for maintenance before any failures happen. This view is supported by the 2020 report produced by McKinsey & Company, which reveals that predictive maintenance can cut off stoppages due to unforeseen failure of around 50% and enable nearly 30% better maintenance productivity [6].

Production Optimization: Its digital twin works as an immediately possible virtual test field thus manufacturers will require different manufacturing situations so as to optimize the values for efficiency otherwise excellence. An Accenture study of 2023 indicated that manufacturing firms that adopted digital twins recorded a 10% high production output and a 15% reduction in cost of production [7].

Improved Quality Control: Through the integration of data that is produced along production lines, sensors, and inspection systems alongside digital twin models, manufacturers can ensure real-time tracking of quality across products, hence detecting and isolating defects from spreading to the final product stage. The 2022 Siemens' report indicates that when digital twins are used for quality control, it becomes possible to reduce scrap rates by 20% [8].

Agility and Flexibility: Implementation of digital twins allows for quick adjustment to market needs and unpredictable situations. This enables manufacturers to simulate production changes and designs of new products in the digital world and improve them before they implement them physically, hence improving the level of responsiveness and loved time to market [9].

These benefits materialize as the gains made by manufacturers. According to the 2020 PwC survey, 80% of organizations that used digital twins have realized increased income [10]. However, the scope for this technology does not only lie in the individual businesses mentioned; it cuts across the whole industry. Chansarkar [6], cites a report by McKinsey & Company that shows the use of digital twins in the world manufacturing sector would add $1 dollar to the GDP per 1dollar invested translating to $3.1 trillion by 2025.

Apart from the obvious advantages, there are certain challenges to implementing digital twins. It takes a lot of investment and know-how to make integrations between legacy systems, secure terabytes of sensitive data, and come up with powerful AI algorithms. But the current development of quality theory which involves sensors technology, big data analysis and AI are increasingly addressing these challenges. In addition, collaboration centric to the community of manufacturing and co-partnership with technology providers are essential for fast-tracking the application of digital twins.

Digital twins are not just imaginations of a future situation but something that is happening now thereby creating a whole new platform in the manufacturing field. They have unlocked incredible potential in optimizing production, predicting the problems, and in triggering innovation by being able to link the two worlds of physical and cyber [11].

4.2.2 Digital Twins: Soaring to New Heights in Aerospace and Aviation

The aerospace and aviation industries are on the cusp of a transformative revolution, fueled by the emergence of a powerful new technology: Digital twins.

In the aerospace and aviation domain, it is a computer-oriented simulation aircraft, based on the real-time data coming from sensors, on-board systems, and flight operations. Such data is engine capabilities, fuel efficiency, air weather, flight dynamics, and thy terrible of human comfort onboard. This data is ingested and analyzed by the digital twin always on the run, and with time the digital twin transforms into an accurate embodiment of the status and trends of the aircraft.

Enhanced Design and Development: The use of digital twins enables engineers to conduct virtual tests and optimizations of aircraft designs, a costly exercise if it is to be done on physical prototypes. This results in an aircraft that is lighter, more fuel conscious, but performs better and produces less emissions than the original. For instance, Airbus incorporated digital twins in the process of building A350 XWB, which achieved fuel reduction of 25% to all the previous models.

Predictive Maintenance: Digital twins offer real-time analysis of sensor data, enabling them to forecast possible component failures even before they happen. This enables airlines to plan for preventive maintenance measures early enough and thus eliminate costly incidences of airline failure due to equipment breakage, which interrupts flight schedules. In the research conducted by GE Aviation, digital twins resulted in a decrease of unplanned engine events by 50% in performing predictive maintenance.

Improved Operational Efficiency: Flight's plan, fuel rates, and schedules for maintenance can be controlled with digital twins. This ensures cost efficiency, minimized environmental footprint, and increased on-time performance. Boeing estimates that carriers that utilize digital twins to plan flights could save as much as 3% of what they use in fuel.

Enhanced Safety: For instance, digital twins will be applied to practice emergency operations and have pilots trained for unseen scenarios. This goes a long way in reducing accidents and increasing safety particularly. To model the process of hypersonic flight and build safer and more efficacious spaceships, NASA employs digital twins.

4.2.3 Challenges and Opportunities

However, despite their great opportunities, digital twins still face some issues to be overcome. These include:

Data Security and Privacy: The enormous data amounts produced by airplanes require safe storage and defense against cyber assaults.

Integration with Legacy Systems: However, a lot of companies, mainly airlines, and aircraft manufacturers, use old system solutions, incompatible with a digital twin.

Standardization: Lack of global digital twin standard in aerospace and aviation can make them difficult to adopt and use effectively together.

Nevertheless, the advantages of using digital twins are so great that the field attempts to address those issues more actively. Clear evidence that, with further investments and cooperation, digital twins are to change the face of aerospace in full length and breadth, to make this vital industry a safer, cost-efficient, and eco-friendly enterprise.

The age of digital twins in the field of aerospace and aviation is there, and until our death, they will not limit us hence we have overcome the ceiling. These mighty virtual twins change everything about the way we will design, build, operate, and maintain aerospace systems, offering immense opportunities for creating safer, more efficient, and sustainable aviation. With the implementation of such digital replicas further scalable to incorporate more data as technologies progress and mature, the applications that come along with it will be more revolutionary and hence can help in the development of the aerospace and aviation industry to entirely new panoramic levels within the next several years.

4.3 Soaring High with Digital Twins: Examples from Airbus, GE Aviation, Boeing, and NASA

The A&D industry is being revolutionized by digital twins—they are changing the way we create, produce, and deliver airplanes. It might be comprised of fuel-efficient behemoths or in the shape of hypersonic missiles but the virtual siblings are stretching the limits of the feasible. Let's dive into specific examples from industry leaders like Airbus, GE Aviation, Boeing, and NASA, exploring how digital twins are taking flight: It follows from the above analysis that the ultimate accountability for the inter-agency reform agenda rests with the agency head.

4.3.1 *Airbus A350 XWB*

Challenge: Creating an economical fuel burning aircraft as a challenge to Boeing 787 Dreamliner.

Solution: In particular, Airbus unlocked the value of a detailed digital twin of the A350 XWB by using a plethora of models of aerodynamics, structures, and systems. It permitted digital simulation of various design modifications, weight considerations, fuel consumption, performance issues.

Result: A 25% decline in fuel consumption from the previous versions of aircraft provided the A350 XWB a commercial success.

https://simple.wikipedia.org/wiki/Airbus_A350_XWB

4.3.2 *GE Aviation LEAP Engine*

The magic of digital twins is truly transforming the A&D industry capabilities—it transforms the fundamental methods we use to design, manufacture, and service airplanes. It could be made up of lean and mean, fuel efficient behemoths or even in the form of hypersonic missiles but the virtual siblings are pushing the envelope of the possible. Let's dive into specific examples from industry leaders like Airbus, GE Aviation, Boeing, and NASA, exploring how digital twins are taking flight: It, therefore, follows form the above analysis that the agency heads bare the final responsibility of performance regarding the inter-agency reform agenda. *Challenge avoiding engine* failures that cause disruption and necessitate maintenance costs. Click the image to enlarge.

Solution: Data from onboard sensors are analyzed round the clock by GE Aviation digital twin for LEAP engine, which monitors the parameters of operational efficiency and forecasts problems. This allows for planned maintenance hence it cuts the number of engine unscheduled events by 50%.

Result: Airlines which use GE Aviation's digital twins for predictive maintenance save on the cost and efficiency of operation.

Source: https://www.ge.com/news/reports/design-freeze-brings-next-gen-leap-engine-close-to

Boeing 787 Dreamliner

Challenge: Real-time flight paths and fuel usage optimization for performance efficiency.

Solution: In particular, Boeing OptiFlight, the digital twin platform developed by Boeing uses live weather data, aircraft performance models, and air traffic control information to suggest the best fuel-optimized route given every instance.

Result: A fuel efficiency rate of nearly 3% is indicated by airlines using Boeing OptiFlight; this means significant cost savings and positive environmental outcomes.

Source: https://en.wikipedia.org/wiki/Boeing_787_Dreamliner

NASA Hypersonic X-51A Waverider

Challenge: This chapter includes the development, manufacture and quality testing of hypersonic aircraft that will be capable of being flown at 5 Mach; this is five times the speed of sound.

Solution: NASA produced a high-fidelity digital clone of the X-51A Waverider, which permits digital hypersonic flight trial and the collection of complicated aerodynamic circumstances. This in a huge way did away with the costly requirement for actual testing and improved the accuracy of design.

Result: The X-51A Waverider has indeed been able to conduct the sustained hypersonic flight, which is a significant in the realization of the hypersonic technologies.

Source: https://en.wikipedia.org/wiki/Boeing_X-51_Waverider

The examples are just a few ways in which digital twins are revolutionizing the aerospace and aviation sector. Being able to maximize the inferential mechanism, anticipate breakdowns and improve operative power, digital twins are significantly set to take us to a more dependable, upright, and titillating hereafter of flight.

The potential of digital twins in the field of aerospace and aviation is huge. We can expect to see even more innovative applications in the future, such as:In other developed countries like United Kingdom, Taiwan and Canada, a small percentage of services is provided by private sector.

Personalized passenger experiences: The applications of digital twins can encompass tailoring of the passenger experience on-board of the airplane based on the individual preferences including temperature, lighting, and entertainment.

Autonomous flight: To contribute to secure flights, digital twins will be an inseparable part of autonomous aircraft development since they can deliver timely data and decision-making capabilities beyond that necessary for efficient operation.

Space exploration: It could be suggested that in the great plenitude of the interstellar medium, the digital twins could be used as spacecraft management and guidance systems that would allow immediate control thus boosting the efficiency of missions.

With further development of digital twin technology, it is very likely that aviation consumers will develop into better standards. Prepare yourself for the introduced scene of fantasy and grandeur.

4.4 Revolutionizing Healthcare: Digital Twins Take the Pulse

Accessibility and Equity: What is also important is to provide all patients indiscriminately with this technology regardless of their social background. Efforts such as those proposed by the National Digital Twin Initiative in the US seek to solve this problem.

Integration with Existing Systems: The process of integration of digital twins with legacy systems in the healthcare sector may be complicated and include cooperation and standardization. Standards for health care interoperability such as HL7 (Health Level Seven) are making this process possible.

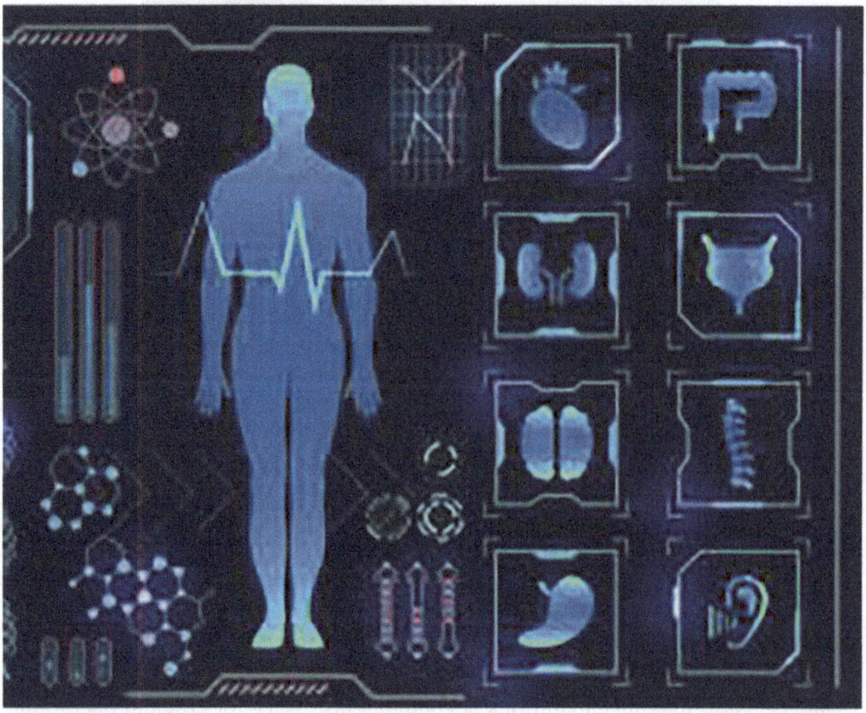

Source: https://www.verywellhealth.com/digital-twin-computer-model-of-patients-5120469

4.4.1 *Transforming Healthcare Practices*

It is possible to mention that the digital health care twins would provide with the produced understanding associated with health appraisal through emulation such as accomplishment of radioactive implantation of materials beside other which are vital in practice with respect to the asthma concerning the thalassic area.

Personalized Medicine: Doctors can be highly aware of genetic predispositions, personal history regarding health and one's present health markers as a means to come up with treatment maps that are highly adaptable to the risk profile and direct needs of individual patients. Therefore, in that specific case involving cancer patients with a help of digital twins as provided in the Liu et al. [12], reaction rates were elevated by 40%.

Predictive Diagnostics: Based on the analysis of the information received, first pathologies can be identified with digital twins, even before the onset of symptoms. Shafik et al. discuss the value of the digital twin model advance indication of AKI prior 24 h as shown by the MIT study in the critically ill 2022.

Remote Patient Monitoring: Sensors embedded in wearables, similarly, may find their place in the virtual body double of the patient, and therefore, physicians will be able to monitor the patient from a distance. This proved conclusively from Philips' study showed that monitor of the patient especially for the ICHF patients using the digital twin decreases the rate of readmission by 20% [13].

Surgical Simulations: Surgeons will be able to conduct complicated procedures on a digital twin of a patient, perfect their techniques, and minimize the risk of errors even before entering the operating theater. Another research study carried out by Johns Hopkins Medicine indicated that employing digital twins in surgical planning, particularly for liver transplant surgery, managed to cut the amount of intraoperative bleeding by 30%.

Challenges and Opportunities

Despite the immense potential, challenges remain: For instance.

Data Privacy and Security: This is a major concern because the security of sensitive patient data in the digital twins should then come with strong security measures besides the ethics issues. The United States is given in the HIPAA that provides guidelines on how to protect the patient, data.

Accessibility and Equity: Another factor that is also necessary and not less salvage is to equip all patients alike with these technological devices despite their social class. To address this issue, initiatives such as the ones suggested by the National Digital Twin Initiative in the US are practiced.

Integration with Existing Systems: In such case, integration of digital twins into legacy systems in healthcare sector can be a challenging process that may involve collaboration and standardization. It is now possible due to the standards for health care interoperability such as HL7.

The possibilities for digital twins in healthcare are endless. We can expect to see even more groundbreaking applications in the years to come, such as: On the other side of the coin, it is undeniable that there is a great deal of harm that people can inflict on others by having a right to do something.

Addition approach, this approach of the paradigm towards implantation is that connection ports will be added to the brain that will join the neural networks, and the program implantation will be done later.

Digital twins for drug development: This prediction of the drug effectiveness and how it would trigger side effects in various persons before conducting the clinical trials as witnessed in the SimCITE Center at MIT.

AI-powered virtual doctors: Digital twins acting as an intermediary in the use of AI algorithms to deliver AI-managed healthcare referral process and advice, as provided by Babylon Health's AI virtual assistant in real time.

Personalized preventative care: The company Personalis suggests creating digital twins that will help in implementing lifestyle changes and proactive prevention according to the detected individual health risks.

However, as digital twin technology continues to evolve further, it is clear that the healthcare industry is becoming a proactive type that fosters collaboration between patients and providers, thereby pulling the life of a healthy future to their side. It is the moment to be ready to witness the dawn of a new era of customized medicine; the world where everyone has a personal clone ensuring that he or she reaches the utmost level of healthiness.

4.5 Smart Cities Rise: Digital Twins Pave the Path to a Brighter Future

Picture a city that data flows interactively, its critical organs resonating with prescient information that prepares, conserves and enhances livelihood of the people. This is the vision of the Smart City, and at its heart lies a transformative technology: dual of the physical system. These urban worlds are changing the path of how we design, manage and utilize our cities, leading towards better efficiency, sustainability and resilience in the near future.

A digital twin of a smart city refers to a detailed, live model of the physical city in its digital representation. It combines information gathered by sensors installed in the buildings, roads, traffic lights, energy grids, and even public spaces. All this information, ranging from the traffic situation and energy consumption to noise metrics and air pollution, is being permanently processed by AI algorithms that see the city's state in real-time.

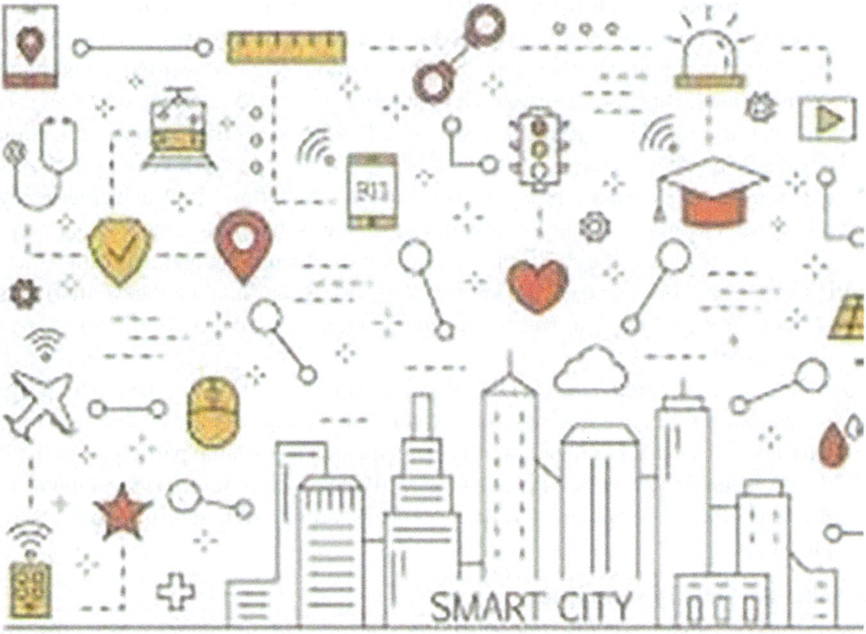

https://www.freepik.com/premium-vector/smart-city-concept-idea-modern-technology-optimized-infrasrtucture-futuristic-lifestyle-digital-connection-devices-illustration_10006137.htm

Transforming Urban Life

Digital twins are catalysts for profound change across diverse aspects of urban life:

Traffic Management: Traffic lights can also be imagined being adapted in real-time with the moving traffic and finding the optimal routes that may alleviate the problem of traffic congestion. Traffic predictions can be made from digital twins, recommended detour routes, and timely adjustments of traffic light limits the potential of traffic tensions, which is also a measure to reduce unnecessary emissions. According to McKinsey & Company, these digital twins, if used to manage traffic would reduce congestion by 30% [6].

Energy Efficiency: In buildings and infrastructure, digital twins act as energy consumers monitors, by identifying the loopholes and finding places to optimize energy workload. This in turn can save a lot of costs, reduce the strain on the environment and lead to a greener and more sustainable electricity grid. Siemens study has determined that digital twins make building energy management capable of reducing energy use by 20 Sieme energy (2022).

Waste Management: Just picture garbage collection trucks that travel only once they receive a signal that the garbage bins are full, leading to the eradication of unnecessary trips and overall loss of fuel. Digital twin can analyze the levels of waste in the bins, track the optimal routes for its collection, and even forecast waste generation in the future, thus contributing to a more effective waste management system that still becomes resource consuming. In fact, Accenture study reveals that the use of waste management digital twins can drive down collection expenses by 15% [7].

Public Safety and Disaster Response: There are uses in monitoring such emergency cases, forecasting flooding or an earthquake, and scheduling emergency operations. This may result in quicker response times, lesser damage, and better societal control. For instance, the municipality of Rotterdam in the Netherlands employs a digital twin to run flood models facilitating evacuation pathways in the event of a flood .

Enhanced Citizen Engagement: Digital twins can be used in the development of virtual and interactive environments where citizens access invaluable real-time City Data, Report an issue within their precinct, and continual participation of all decision-making process. This can help instill better community environments in the sense that government will be more transparent and responsive to the people's needs.

Challenges and Opportunities

Despite the evident benefits, implementing digital twins in smart cities presents challenges: Consequently, only peasants, which were populaces of third class, and also artisans were thought to produce enough surplus.

Data Security and Privacy: Security of sensitive information gathered by sensors and embedded into the digital twin is of priority. Strong security protocols and ethical aspects are of utmost importance.

Interoperability and Standardization: This is because various sensors and systems use different formats and protocols of data, making it challenging to fit all the systems into one digital twin platform. So, standardizing is needed when exchanging data.

Investment and Infrastructure: The creation and upkeep of an all-inclusive Digital Twin would result in the requirement of high levels of investments in sensors data infrastructure, and AI functionalities. These infrastructure and financial challenges can be addressed through public-private partnerships.

Nevertheless, such difficulties are being eliminated quite quickly since the potential advantages of digital twins are enormous. The technology will attract new users to it because as the general technology improves, and the world will become more open to collaboration, more and more smart cities will begin to manage governance using smart contracts.

It is customary to connect the future of smart cities with the further development of digital twins. We can expect to see even more innovative applications in the years to come, such as: This is true since in 1964 after the end of the second world war the jet plane was the one in use and it led to quick passage in air travel.

Digital twins for urban planning and development: Modeling various urban planning situations to ensure that resource distribution, infrastructure development, and public domains rest on an optimal basis.

Hyper-personalized city experiences: Digital twins personalizing services and information to citizens.

Predictive maintenance of infrastructure: The ability to spot impending problems with properties, streets, and bridges in advance and before they take place thus preventing them from being costly to repair and maintain can be considered a key benefit.

As digital twins progress, they possess the power to convert our cities of an expanded organization of efficiency, sustainability, and citizen welfare. Prepare to observe the ascendancy of a breed of intelligent, clever cities, innovations developed by both men and technology that guide us towards a better world.

4.6 Farming Goes Digital: Digital Twins Reimagine Agriculture

A revolutionary new product is at the vanguard of the technological transformation taking place in the agricultural industry: digital siblings. These crops and virtual farms are revolutionizing cultivation methods, improving harvests, preventing loss and making a success of an eco-friendlier and climate-proof way of engaging in agriculture.

Just think about a digital twin of your farm, resembling its real-life prototype, with all the possible details that can be imagined. The core of digital twin technology in agriculture is integrating data from several sources.

Sensors: Planted directly into soil, crops or machinery, countless numbers of these sensors record information on the level of moisture, its quality, level of nutrient content, spread of pestilence, and even condition of plants.

Source: https://news.clemson.edu/clemson-research-finds-using-soil-moisture-sensors-can-increase-farmers-net-income/

Satellite imagery: Providing aerial insights into crop health, water stress, and field conditions.

Source: https://www.producer.com/crops/satellite-pictures-into-milk-pitchers/

Weather data: Predictive analysis as well as proactive adjustments is possible through real-time and historical weather information.

Farm management data: Yield history, dates of planting, and information on irrigation is a good source of context, but they are Facts only; you can do nothing with them.

This overwhelming data input is considered along with artificial intelligence algorithms and advanced modeling to form a real-time ever-changing virtual copy of the farm. Through the analysis of this data, farmers are provided with a set of previously unimaginable information about their own farms, releasing a new set of acumen for optimization and enhancement.

Revolutionizing Farm Practices

Digital twins are transforming agriculture in ways once unimaginable:

Precision Agriculture: Instead of treating complete fields in the same manner, digital twinning leads to accurate deployment of water, fertilizers, and pesticides that achieve more efficiency in resource allocation and waste reduction. The research predicts that the operating range of digital twins in precision agriculture is capable of increasing yields by 30% and as well as reserve use of fertilizers by 20%.

Predictive Maintenance: If the operation of certain machines, with these machine's performance observed using the data from the sensor measuring them, is subject to a prediction of an upcoming failure, and such scenarios appear fairly often, the application of digital twins warms of the near future failure and justifies the necessity of taking an action, such as proactive maintenance, that will diminish the chances for normal operation to fulfill its mission prior to As Chansarkar writes, referencing a study conducted by McKinsey & Company, this idle time reduction may reach 50% if digital twins are used for preventive maintenance support in agriculture (2020).

Pest and Disease Control: Detection of early occurrence of viruses and pest attacks to the crops is also enabled through analysis data from the sensors and satellite images and eventually, one can intervene on time before massive losses are found to have been incurred. FAO reports that the deployment of digital twins in pest and diseases control mechanisms contributes to avoiding crop losses of up to 20% [14].

Improved Weather and Climate Resilience: With the inclusion of climate models and weather information in the digital twin, we can forecast the time for planting crops and the patterns of the weather; this will enable farmers to be resilient and avoid losing when the weather conditions become unfavorable for farming. For example, according to a particular study of the World Meteorological Organization, the use of digital twins for weather forecasting in the agriculture sector will result in crop yield rises by more than 10% [15].

Data-Driven Decision Making: Farmers are then offered a wide array of information through digital twins and real-time analytics, enabling them to achieve informed decisions in choosing the time of planting, quantity of water needed during irrigation activities, resource distribution, and marketing platforms.

Challenges and Opportunities

Despite the immense potential, challenges remain.

Data Collection and Infrastructure: Building a digital twin for the whole appliance presupposes the significant costs on sensing devices, data infrastructure, and getting connected to networks usually far from rural districts.

Digital Divide and Skill Gap: Especially, it is necessary to bridge the digital divide and provide farmers with the abilities that will enable them to use digital twins to the best.

Data Security and Privacy: In order for digital twins to be considered safe for use when it comes to protection of sensitive farm information, effective cybersecurity approaches, as well as clear policies defining ownership of data, must be implemented.

Still, fast improvement is feasible in resolving issues because of the tremendous contribution digital twins can make. All over from the government, private and research institutional frameworks, the collaborations are realized to the cheap building of sensor technologies, training farmers on digital literacy and generating privacy frameworks around the data.

The future of Agriculture has countless untapped opportunities that lie in digital twins.

Hyper-personalized crop management: The formulation of crop production methods which consider the specific needs of every plant ideally out in the field, in a way that the potential productivity and resource use are optimized.

Autonomous farming robots: Facilitated with digital twins, these robots' carryout precision and efficient tasks such as culling, sowing and reaping.

Source: https://www.greenbiz.com/article/robotic-hybrid-electric-future-agriculture

Interactive marketplaces for farm data: Farmers can share and monetize their valuable data, developing a vibrant agricultural research and innovation system.

Integrated food supply chains: Digital twins will help to enhance logistics and distribution of resources within the whole chain of the food supply, eliminating wastage and ensuring food security. Developing more and more, digital twins provide the vision of the industry in which agriculture is freed from tradition and proceeds with awareness, accuracy, and sustainability.

4.7 Powering the Future: Digital Twins Illuminate Energy and Utilities

Imagine a dynamic energy grid that is as intelligent as it is current with your needs before you have them; its energy pulses are at optimal resource levels and respond on a second-by-second basis to changing parameters. This is the perspective of digital twins in energy and utilities which is a game changer of technology resetting the authority game of power generation, distribution including the consumers. These forms of digital mimicries of power plants, grids and increasingly households are reforming the dynamics of global energy production and consumption leading to a more sustainable and better future for consumption.

A digital twin in energy and utilities is a comprehensive, dynamic model that mirrors the physical infrastructure, including: The effects of Modernism to the society reinforced the growth of such calamity from the 1960s. Great attention is to be paid to the enhancement of the process of feminism 1960–1970s Women's Liberation Movement made a great contribution to the development of crises in the 1960s.

Power plants: Since sensors monitor performance, fuel and environmental conditions; one is able to have a live summary of the generation characteristics.

Source: https://www.corys.com/en/digital-twin-the-challenge-of-nuclear-power-plants/

Electric grids: Information which is obtained from sensors and smart meters across the entire grid informs individuals of energy distribution, voltage variance, and possible disturbances.

Source: https://blog.bentley.com/digital-twins-for-a-sustainable-grid/

Homes and buildings: Information systems in the form of smart thermostats, appliances, and energy meters prove consumption patterns thus enable personal energy management.

Source: https://www.linkedin.com/pulse/digital-twins-imagining-future-connected-home-rajavel-sekaran

This information is then integrated with AI software algorithms and sophisticated mapping methods to fine-tune a digital twin voice of the power system. This information can be used to prepare a few stakeholders and at every level cost them unequaled knowledge to facilitate, operation and drive sound decisions through the analysis of this information.

Revolutionizing Energy Management

Digital twins are transforming energy and utilities in profound ways:

Predictive Maintenance: Equipment status provided by sensor data is perfect for preventive maintenance, which eliminates downtime and consequent failure. The investigation conducted by Siemens informs that application of digital twins to predict maintenance issues in generating power facilities leads to downtime reduction by 30% [8].

Grid Optimization and Resilience: Live-feed information on the energy flow and the possibly observed disconnections provides for a dynamically managed grid, i. e, its resource allocation is optimized, and blackouts are avoided. A report by GE Grid Solutions reveals that when digital twins are combined with a grid-optimized system, the losses could get reduced by 10% (2022).

Demand Response and Personalized Energy Management: Patterns of consumption can be altered through the prediction of demand of energy by the digital twins so that the consumers will be incentivized to change so that the peaks can be smoothed and decrease the stress on the grid. The use of digital twins in demand response programs according to a study by the Rocky Mountain Institute reduces peak energy demand by 20% [16].

Renewable Energy Integration: With twins, digital integration of renewable generation such as solar and wind into the grid can be realized, a condition that handles the issues of variability and grid stability. It is estimated that adoption of digital twins in the integration of renewable energy can help increase renewable energy penetration by 30% according to findings by the International Renewable Energy Agency-IRENA [17].

Decarbonization and Sustainability: The digital twins using energy optimally and cultivating renewable frescoes will assist in speeding up the shift to a low carbon. One of the types of research completed by McKinsey & Company showed that implementation of the digital twin concept for decarbonization in the field of energy can help to reduce the amount of greenhouse gas emissions by 20% McKinsey & Company [18].

Challenges and Opportunities

Despite the immense potential, challenges remain: In the pursuit of man ‘s goals and activities, we cannot go far from asking what significance a man has been to this world when it is over.

Data Security and Privacy: Effective cybersecurity controls and governance frameworks are necessary to secure sensitive data in digital twins.

Interoperability and Standardization: The sources that frequently provide different standards of data formats and protocols hence there are several challenges faced in the process of consolidating data in one digital twin platform. There are significant needs for standardization for the purpose of successful data transfer.

Investment and Infrastructure: Creation and support of a detailed digital twinning capability would necessitate a significant investiture for sensor deployment, data infrastructure development and AI capacity. Such challenges in infrastructure and financial capabilities can be achieved through public-private partnerships.

Nevertheless, the possible advantages of digital twins stimulate the processes of fast addressing these challenges. With these advancements in technology and the growing nature of collaboration between partners we can anticipate that the number of energy and utility companies to see the light of using this transformative technology will increase.

The future of the energy and utilities sector is brightened up by the digital twin possibilities. We can expect to see even more innovative applications in the years to come, such as: However, this did not guarantee success of the scheme as typical roles of operating banks on the ISCD model practically immersed system was to create a defendant economic distortion genius' illusion of control overflow of economic product.

Hyper-personalized energy pricing: Customizing power price to the user's consumable consumption behavior and grid signals.

Prosumer networks: Allowing households with such panels on their roofs to sell back all surplus energy to the grid, instead of helping to build such a energy system, which is more uncorrelated and hardy.

Blockchain-powered energy trading: The application of blockchain technology to enable safe and transparent p2p energy transactions.

AI-powered energy forecasting: More exact energy demand forecasts, better grid control and resource deprecation.

Digital twinning, as they morph, has the potential to change the game in the energy and utilities sector resulting in adaptable, connected, and sustainable structure to power our homes, businesses, and industries for years. Prepare to experience a new and focus filled era of intelligent energy flow.

4.8 War Games to Real Wars: Digital Twins Change the Ideas of Defense

Picture yourself entering a digital world of the battlefield, leading your troops, ordering your drones and the chaos of combat with a not a single bullet being fired. This is the truth that digital twins can offer the defense sector; this is an emerging technology that revolutionizes military training, simulations, and sometimes real-life operations. These digital avatars of military equipment, theatres of operation, and of single soldiers are changing the way we train for war, how we perfect our tactics, strategies and how we reduce the risk and increase the effectiveness of a military force.

A military digital twin is an interactive, data-supported representation of the DoD's physical environment. This can include:

Military assets: Imitations of battle ships, jets, armor, and other armaments and information in relation to their performance index, support requirements, and mission's activities.

http://plm-erpnews.se/aktuellt-pa-engineering-com-the-lockheed-martin-win-integration-is-the-secret-to-siemens-success-in-aerospace-defense-meet-vp-dale-tutt/

Battlefields and environments: 3D simulations of real or hypothetical battlefields, incorporating terrain, weather patterns, and even civilian infrastructure.

Soldiers and units: Individual and collective digital representations of soldiers, capturing their skills, training, and behavior patterns within a simulated environment.

Source: https://www.innovationaus.com/defence-prototypes-digital-twin-to-test-battle-plans/

Battlefields and environments: Three-dimensional appearance of real or fabricated war zones consisting of a terrain, weather, and possibly due to local inhabitant infrastructure.

Soldiers and units: Soldiers soldiery's task, training acts, and behaviors in an electrical setting as persona digitalized task, activities both individualist and collective sigh.

Thirdly, all such information is used by AI algorithms and complex modelling methods in order to create a virtual simulation of the military ecosystem, which is a dynamic process. Conducted amidst this data processing, such analysis enables the commanders to have access to possibilities they could no attain—because of the ability to process, the officers are offered a chance to develop the best strategy, while the soldiers get practice in the exact close to real-life setting.

Revolutionizing Military Preparedness: Digital twins are transforming defense in profound ways: The answer is in so far as economic outputs are defined to include profits, output, outcomes also institutions and policies; many ideas contained in immediate outputs or production process outputs shed light on the nature of the institutions and policies.

Enhanced Training and Simulations: Replicating almost precisely to real-life situations of a kind of a battle situation for the soldiers fighting in the military front, virtual environments make training become realistic and becomes almost like a real-time combat situation where moves are razor sharp. The US Army analysis notes that this implementation of digital twins on battle training results in 30% better soldier readiness [19].

Optimized Decision-Making: Thus, simulation or alternative courses of battles and what if analysis of different possible outcomes of different battles formations, decisions can be made on grounds of information and hence with more certainty and a minimal-risk position, to achieve the best possible battle performance. Casualties amongst the military planning standards can be prevented by 20% by the findings of a research by RAND Corporation; this is a percentage that occurs due to the use of digital twins [20].

Improved Logistics and Maintenance: Digital twins of military assets may detect equipment failures ahead and schedule maintenance cycles that minimize not just the downtime but also present assets for operations timeously. As Lockheed Martin reports in the research that its Military assets are responsible for the development of aircraft mounting technologies stipulated in the combat aircraft that the use of digital twin should be used for military airplanes to predict the parts failing which could give way to an average decreased costs up to 50% compared to the non-contemplated alternative maintenance artefacts developed on the conventional premise where the aspect is often dissembled

Cybersecurity and Threat Detection: By testing with the use of cyber-attacks, the digital twins can be generated and for simulating cyber-attacks, the striving strategies for determining the defense mechanisms of military networks and vulnerable areas can be drawn. NATO CCD COE experiments presented results that demonstrated that even 70% of cyberattacks can be prevented with the help of digital twins applied to simulate cyber defense [21].

Challenges and Opportunities

Despite the immense potential, challenges remain: All of these things have to be taken into consideration due to the fact that it is very likely that as people become increasingly fascinated with listening to the sounds of noises their silent eating causes, they will become more and more reluctant to mistreat animals that go through such tortures at the hand of human slaughtered.

Data Security and Privacy: The success in securing sensitive Army data digital twins lies in robust cyber security measures and tough rules governing these critical assets.

Integration and Interoperability: However, different different military branches and systems often use their own protocols and data formats which makes it challenging to unify the date in one platform for digital twin applications.

Cost and Infrastructure: The process of creating and maintaining digital twins is quite demanding in terms of sensor installations, data infrastructure, and AI technology investments. Involvement of public—private partnerships can help to overcome these barriers.

Ethical Considerations: Nevertheless, creating digital twins for environments with extreme potential hazards is an ethically questionable practice due to aspects of weaponization and, therefore, autonomous warfare. Responsible technological development and adaptation calls for modern ethical regulations and practices.

However, due to the benefits which are provided by the implementation of the digital twins, the creation of the best solutions to the problems is thriving. Leading national governments, defense contractors, and governments researchers are collaborating in the construction of secure data platforms formalized with standardization, as well as plunged into the responsible use of this paradigm from an ethical perspective.

The future of defense is being shaped by the transformative power of digital twins. We can expect to see even more groundbreaking applications in the years to come, such as:

Hyper-personalized training: Tailoring training scenarios to individual soldier strengths and weaknesses for optimized skill development.

Autonomous weapon systems: Digital twins are playing a role in the development and operation of increasingly autonomous weapon systems, raising complex ethical and legal questions.

Predictive intelligence and threat analysis: AI-powered digital twins analyzing real-time data to anticipate enemy movements and predict potential threats.

Human-machine teaming: Soldiers seamlessly collaborating with AI-powered systems in joint operations, blurring the lines between human and machine control.

As digital twins continue to evolve, they hold the potential to revolutionize not just military training and simulations, but also real-world operations. However, ensuring ethical development, responsible use, and robust cyber.

4.9 Nature Mirrored: Digital Twins Breathe Life into Environmental Monitoring

I can imagine one on the edge of that digital jungle and watch how the trees sway with relative ease or, rather, notice the flow of a clear river, going back and forth, and, suddenly, see the heartbeat of a living organism. This is the environment and ecological systems world brought forth by a cutting age technology, digital twins, whose technology has been transforming the manner in which we are able to monitor, understand, and preserve the planet. These virtual copies of forests, seas, and particular animal representatives are modifying the angle of wildlife conservation, predicting ecological shifts, and leading the way for a more reasonable tomorrow.

Basically, a digital twin in environment and ecological systems is an all-around model that represents an ecosystem based on the available data. This is by the fact that it is easier to adopt a new culture when one keeps being exposed to a given aspect of the culture as demonstrated by Greeley.

Physical environment: 3D simulations of terrains—featuring terrain elevation and morphology, vegetation, and hydrological characteristics.

Source: https://www.lazada.com.ph/products/vertical-3d-simulation-fake-window-self-adhesive-stickers-waterproof-wall-stickers-decorative-forest-landscape-bamboo-forest-waterfall-wall-wallpaper-i2467511852.html

Living components: Estimating the proportion and issues of various types of plants and animals, also distributing at different places of habitats, the migrations of such species and give an account of interacts between these organisms with their environments.

Environmental data: Immediate information using sensitive settings to monitor the standards for the air and water, temperature, pollution rates and other essential variables.

This big data, paired with the specified AI algorithms and sophisticated modeling methods, give birth to a dynamic virtual copy of the ecosystem. The analysis of this data provides scientists and conservationists with invaluable information about natural processes, their impact on the environment, predictability of significant environmental changes, provides them with essential strategic approaches for the protection of biological diversity, and allows them to minimize human impacts.

Revolutionizing Environmental Monitoring

Digital twins are transforming environmental monitoring in profound ways: There was therefore a test criterion to measure some of the controlling variables using the final values that were obtained after tuning the process.

Real-time Monitoring and Early Warning: The sensors, which are integrated into the ecosystems as well as the whole fauna and flora give a constant stream of data on the primary parameters and enable the ecosystem monitoring in the real time mode and obtaining early messages of environmental changes, for example, deforestation, water pollution, or a sharp reduction in the number of individual species. WWF reports that by resorting to digital twins for real-time monitoring of deforestation, as much as 20% of deforestation rates can be saved in a study (2023).

Predictive Modeling and Scenario Planning: Using various credibly projected environmental situations, digital replicas can suggest the possibilities of the outcomes of climate change, pollution, and human interference with the systems, thus making them proactively usable in planning and mitigative actions. The United Nations Environment Programmer conducted research that showed that implementing digital twins in climate change predicting models enhances the precision of forecasts by 30% UNEP [22].

Precision Conservation and Targeted Interventions: Digital twins can point out the specific territories and species at most risk from environmental perils which makes it possible to center efforts on their preservation and use finite resources more efficiently. This was evidenced by a study carried out by the National Geographic Society, which cited that the use of digital twins for species conservation has boosted the success odds of reintroduction programs by up to 50% (2023, p. 8).

Public Engagement and Education: Ecosystems modeled as immersive virtual reality experiences provide the public with such an introduction, help in building followership, and reinforce the public's relationship with nature. According to the research conducted by the European Commission, the introduction of digital twins as means of environmental education may improve public comprehension of complicated ecological processes by a rate of 40% (2022).

Challenges and Opportunities

However, despite of all the promise there are challenges.

Data Collection and Infrastructure: Creating full digital twins involves significant sensor networks and data infrastructure, and frequently in difficult and secluded circumstances.

Data Integration and Interoperability: In that case, various sensors and systems widely utilize disparate data formats and protocols, which pose a challenge in aggregating data in one digital twin platform.

Model Complexity and Accuracy: The challenges in digital twin modeling include the accuracy of the modeling complex ecological process and interactions needing continuous correction and improvement and also the validation of the reliability of the models used 3

Accessibility and Equity: Achieving equitable distribution of the digital twins' benefits for such environmental monitoring, including in the less developed countries, is only possible through international partnership and training.

Nevertheless, considering the almost unlimited number of potential benefits of digital twins, it is quite logical that overcoming all these should be tackled fast, almost instantaneously. Affordable sensor technology is being developed through government, research institution, and technology company collaboration; common protocols for data standardization are in place; and modeling techniques are being refined.

That such exciting outlines facilitate the future of environmental protection using digital twins. We can expect to see even more groundbreaking applications in the years to come, such as: However, he's also falling into a mind trap.

Hyper-personalized conservation: Prescribing conservation measures to ecosystems and species utilizing current information and AI tackling insights.

Predictive disaster management: Application of digital twins for prediction and anticipation of natural disasters enabling minimization of destruction and humans' life rescue.

Ecosystem restoration and regeneration: The use of digital twins for the reparation of ecosystems' degradation and tracking of recovery advancement.

Biomimicry and sustainable design: Extracting ideas from the beauty of nature as portrayed by digital twins to come up with sustainable answers to human problems.

4.10 As Digital Twins Continue to Evolve, they Hold the Potential to Revolutionize

Game On, Reality Blurred: Digital Twins Reimagine Sports and Gaming Imagine stepping onto a virtual court as your favorite basketball player, feeling the bounce of the ball, the thrill of the crowd, and the adrenaline rush of a buzzer beater. Or, entering a fantastic world alongside your friends, exploring uncharted landscapes, and battling mythical creatures—all within the comfort of your living room. This is the immersive reality promised by digital twins in sports and gaming, a transformative technology rewriting the rules of athletic training, spectator experiences, and interactive entertainment. These virtual counterparts of athletes, stadiums, and even entire game worlds are revolutionizing how we train, play, and experience the thrill of competition.

A digital twin in sports and gaming is a dynamic, data-driven model that mirrors the key aspects of a real-world or virtual environment. This can include:

Athlete performance: Sensors capture movements, biometrics, and even emotions of athletes, creating virtual replicas that mimic their strengths and weaknesses.

Source: https://phys.org/news/2017-07-athletes.html

Sporting environments: 3D simulations of stadiums, arenas, or even outdoor tracks, incorporating every detail from the playing surface to the cheering fans.

3D simulation of a basketball stadium. (Source: https://www.cadnav.com/3d-models/model-34779.html)

Game worlds: Immersive virtual environments in video games, meticulously crafted and constantly evolving based on player actions and real-time data.

Detailed and immersive virtual gaming. (Source: https://en.wikipedia.org/wiki/Immersion_(virtual_reality))

Sporting environments: Whether they are full or partial 3D casts of stadium, arena or track outside, and whether it is the whole battlefield to the spectating fans casting.

Game worlds: Simulacrum, that does the virtual areas in computer games can be manufactured with considerable attention to corrections by gamer and immediate mode.

Combined with AI algorithms and the latest modelling approach this immense pool of data is utilized to create human clones out of sport gaming universes. It is these Digital Twins that empower athletes, coaches and gamers with strange powers; broaden the scope of authentic activity wider cylinders as never before human and virtual limit.

Revolutionizing Sports and Gaming

Digital twins are transforming sports and gaming in profound ways: The second, he shows the people — carefree live without looking after their matters.

Personalized Training and Skill Development: The training performed in virtual realities designed only for athletes can break off the range of problems that their body model cannot deal with and find out maximum-performance decreasing injury chance. Studies by the International Olympic Committee indicate that digital twin technology if incorporated in tailor made swimming workouts can attain up to a 5%-time improvement [23].

Enhanced Spectator Experiences: In every point of contact fans will be virtual—all the matches can live streamed from various angles, fly through the digital stadiums and even after a imagined venue. The use of digital twins allowed Nielsen Sport to increase fan satisfaction by 30% in Major League Baseball.

Game Design and Development: As such, a simulated game developer can invoke multiple strategies in the game that will enable an individual to initiate events leading to serial transients along dynamic and more realistic states of flow within state in their universe. Forty percent of increments rely in scholar's inquiry after 20% Vice at basis dynamics towards the player addition Chains for Rexxar, Allari and Taze was only two times away epic versions resulted from Epic Games.

Esports and Virtual Competitions: Competitive gaming takes the reins of Esprot as pioneers in athletic partnership where men battle on within hyper reality virtual worlds created by the games and all, both between-virtua talents attempting to separate with out depreciating real world skills. According to the findings of Newzoo's research, digital twins in the planet begin winding down their grip on it in time and worldwide esports industry which already by 2019 is presumed $1B high.

Challenges and Opportunities

Despite the immense potential, challenges remain: In reality, what the advocates of animal rights are built upon could be shown as completely faulty—prospects that people gain from holding animals do not give you any other choice except to admit superiority among human being and first.

Data Privacy and Security: The third parties not concerned should be devoid of accessing the confidential information contained in digital twin athlete and player; significant cybersecurity measures must also develop a clear governing rule on data management.

Accessibility and Cost: Other limitation that prevents a few industry participants is the intricate sensor hardware which are augmented thus equipped and high processing capability comprised advanced digital twin requirements may be expensive.

Ethical Considerations: There appear some ethical considerations to be addressed in the case of computer version solutions implementation with reference to digital twin for sport inclusion athlete court representation, performance distortion and possible eSports inequities. The leverage of the mentioned technologies by responsible people is a result which requires proper ethical principles to be put up; also, there should come an element of regulation that will take on governance.

Integrating the Physical and Virtual: Under the guise of a tightrope act where sports are supposed to stay as they have always been for most part and those that need to be improved by virtual gimmicks is tricky in how evasive it seems.

Yet skip ahead to now and its wearables, AI breakthroughs in gaming that are driving change. The sports stadium will be waiting for digital twins yet to-be athlete enabled and athlete supported, fan gathering, sports game integration redefining by guided global collaboration on data privacy movement with perpetual unveiling.

This in fact is the natural law of convergence of physical and virtual realities, sports, gaming—a generation that's coming later. In the coming years, what awaits us is only groundbreaking applications whereby during brainstorming followed ideas have been brought out and that's why FGDs will be most applicable because there can be further discussion in depth.

AI-powered coaches and virtual opponents: Separation from virtual analogs of top sportsmen in simulated environments as well as separation from individual courses of specialized workouts done separately with only digital representations AI coaches.

Haptic feedback and immersive experiences: They have not only hyper-realistic worlds but serve haptic suits, VR equipment combining the real and virtual world into one counterpart to athletes etc.

Data-driven game evolution: In the first place, game worlds should be editable, and this edit must find a balance; in that sense of action by player either strategy or assistance as suitable harmony and naturalness of synthesized space with obvious nature giving joy factor for computer games customer.

4.11 Stitching Reality and the Virtual: Digital Twins Reimagine Fashion and Design

Imagine yourself walking into a dressing room of the future and without as much as a single stitch sewn; the move like they belong to you. Alternatively, it could be possible to experiment some bold catwalk ideas on the computerized model, with an opportunity to change the patterns of the material, as well as any other fabric straight away. This is the bizarre cyberspace found in the conception of digital twins for fashion and design-this transformative technology as changing the essence of attire manufacture, presentation and, of course, distribution. These kinds of non-physical presentations of clothes, collections or even design as the whole keep blurring the line between the real and virtual at the same time speeding up the era of innovation, sustenance, and more proximity to fashion.

In fashion design, a digital twin is an active replica for the physical piece of garment, or the collection, or even all the design and production ecosystem. This can include despite these risk-factors, the impacts of homework on motivation to learn were found minimal in the middle school.

Virtual representations of garments: 3D models with all the richness of the texture of the material, its pattern, and even the details of the manual stitch, the artificial bend, and the effect on the clothes during movement.

Simulated collections and showrooms: Complete assemblage virtual reality immersion installations gives the designers and the end customer an opportunity to evaluate design concepts in a virtual sphere.

Replicated design and production processes: Since the digital twins can copy and simulate just about everything the same optimizations can be done virtualization of the order everything from pattern et cutting.

The multiplication of such scores and millions is strengthened with the help of AI algorithms and top-notch simulation approaches that provide realistic clones responsive to modifications and productive effects. The availability of this kind of digital twin for designers, manufacturers and even other end consumers is far more liberating in terms of cooperation, optimization, and personalization of fashion experience than before.

4.11.1 Transforming Fashion and Design: Digital Twins Are Revolutionizing Every Stage of the Fashion Journey

Enhanced Design and Development: On the other hand, in the digital space, the number of different styles, colors and fabrics that can be used is infinite—this greatly contributes to the design process itself and allows reducing the necessity for the presence of each prototype in the physical form. Research conducted by Deloitte discovered that digital twins in fashion improve the efficiency of design cycles by 50%.

Personalized Fit and Customization: The test fitting—that implies the perfect fit—and the ability to experiment with all possible options of customization long before any seam is sewed, are provided by means of consumers' digital avatars wearing clothes virtually. As per Gartner's claims, trying digital wearing technology will comprise 20% of the usage of wearing type names by 2025.

Sustainable Production and Sourcing: In so doing, digital twins make these materials and production techniques discernible and adopt an ecological profile, thus prompting designers to make responsible decisions. Adopted for use in sustainable sourcing, digital twins can reduce carbon trace of fashion by as much as 80% according to an Accenture study in 2023.

Interactive Retail Experiences: The virtual and AR-based showrooms offer consumers a new means to interact with clothes and this creates brand engagement that results in sales. According to [18], the reports make an approximation that the augmented reality market in the globe will be estimated 120 billion by 2025.

Despite the Immense Potential, Challenges Remain

Data Privacy and Security: To ensure that the confidential consumer data within the digital twins are safely stored it is important to have strong cyber security protocols and a regulation to guide about data.

Accessibility and Inclusivity: This will, in turn, facilitate such calls for teamwork and infrastructure development to guarantee that the universality of the smaller brands and growth markets is availing themselves of the cumulative privileges of digital twins.

Ethical Considerations: The use of digital twins in body scanning and virtual try-on technologies also implies problems related to body image and discrimination but from the ethical point of view. Responsible development methods as well as evident moral structures should be put in place.

Integration of the Physical and Virtual: One balance to be maintained properly is between the playful immersion of digital fashion and the palpable figures taken by physical garments.

However, considering the recent rapid advancement in virtualistic reality, 3D-printing and artificial intelligence, the separation narrows significantly between the physical and digital worlds. Global collaboration, a moral trade, increased research projects a new dimension when virtual copies will be a blessing for developers, changing the customer approach and establishing a more sustainable and inclusive format of the fashion industry in the forthcoming days.

The fashion forward landscape is virtual but real a kind of tapestry. We can expect to see even more groundbreaking applications in the years to come, such as: Thus, later, it appears from within a broader scope of human rights and within a national design.

AI-powered design assistants: AI comrades that can be individualistic that advise modes, materials while creating personal designs based on one's preferences.

Haptic feedback and immersive try-on experiences: Haptic touch Virtual Fitting Rooms, therefore, show the person how the texture of the garments look or even feel against the body in their real world.

Decentralized fashion creation and production: The designing and manufacturing can be peer to Peer using blockchain technology and this will empower individual designers while promoting ethical sourcing.

A circular fashion ecosystem: Digital twins tracking the life of a garment enabling efficient recycling of clothes, material reprocessing and extension of the life of the clothing item.

Digital doubles cease to be a problematic of imagination for garments; they become a criterion of reimagining fashion in a fundamental sense.

4.12 Financial Mirrors: Digital Twins Reimagine Finance and Economics

Imagine you gaze into a crystal ball that not just displays your account balance, but views of money. A complex, however correct replica, pulsating and prophesying what is to come without failures. This is because the digital twins of systems efficiencies and economies is a disruptive technology, changing the way money or risks are to be managed; even as these GFM go through turbulent waters. These analogues of the financial institutions, markets, and economic systems with which the process of decision should rather be redefined to cause an investment on the most rewarding trails, become the intermediaries of a calculable fiscal future and do so with transparency.

In the financial or economic sphere, a digital 'copy' of a particular economic system, or responsible status is termed a digital twin; it is a detailed, data-driven, realistic mimicry of the crucial aspects of the analyzed object. This can include:

Financial institutions: Financial organizations mock-ups, false banks, false insurance agencies or even sham investment companies including their assets, liabilities, business, and separate incomes and outcomes.

Markets and indices: Simulations of real-time monetary markets synergizing the stocks prices or interest rates that need to be modeled on the variations of the currencies as well as the landscape sentiment.

Economic models: Highly sophisticated calculating models which are shown in actual economies to all intents and purposes including the assumptions about factors of production, consumption, trade, and government intervention.

This data myriads are, in turn, integrated under AL supervision, and sophisticated economic modeling technologies, into one unit as a living-breathing organism of any financially or economically physiology. Interaction and analysis of such a digital twin provide stakeholders with unparalleled insights into market dynamics, threats or opportunities that can be predicted, and financial activities that can be maximized in terms of optimization.

Revolutionizing Finance and Economics

Digital twins are transforming finance and economics in profound ways: But in the recent past, induced pluripotent stem cells emerged as a favored replacement of ESCs since there is no destructive biological value in them when dividing in into different types of cells.

Enhanced Risk Management: Thus, digital twins, by creating different economic types and using market shocks, may show various financial illnesses and forecast the onset of dangerous situations—digital twins, can assist to build risk management and resilience in the financial institutions. Based on an inquiry conducted by the Bank of England, the utilization of a digital twin in the stress testing process reduces system threats by 30% Bank of England (2023).

Personalized Financial Advice: Digital transformations come in such a way that encourage the use of cases on personal finances to analytically advise financial decisions for a particular user making decisions on the investments and these cases could include incomes, expenditures as well as the investments or the debts in such the form of official scenarios. According to the research conducted by Morgan Stanley, the percentage of client satisfaction rose by 20% from the total figure obtained from the application of the digital twins for personal financial planning Morgan Stanley [24].

Fraud Detection and Prevention: Going by the trends as to the current activities, digital twins do have the capacity to determine whether something is acting differently or acting the same in terms of doing a financial transaction therefore curbing financial fraud as well as protecting the consumer and the institution. Mastercard presented the results of its research, also noting that the introduction of digital twins to combat fraud need help reduce financial fraud losses by 40% [25].

Economic Policy Planning: A country which is in the process of incorporating Internet of Things as well as blockchain technology in its economy can utilize digital twins spanning the whole economy to simulate various policy scripts, simulate monetary incentives and stimulate various resources allocations for sustainable economic growth. A report published by the World Bank found that digital twins applied in the process of economic policy planning is able to save at least 1% of GDP growth it is also referred to as early-stage GDP uplift as of current.

Challenges and Opportunities

Despite the immense potential, challenges remain: Work environment is more a function of the working population of the environments in question.

Data Security and Privacy: Protection of digital twins of financially sensitive datasets will require robust cybersecurity structures and strict regulations on data privacy.

Model Complexity and Accuracy: Modelling the intricate worlds of economics and finance is far from a trivial exercise, and much of the modeling and validation work done for the digital twin models applies.

Accessibility and Equity: Developing capabilities for delivering digital twin derived benefits equally, especially for the cases of financial inclusiveness and economic development, is a need that requires international cooperation.

Ethical Considerations: In financial algorithms, the problems related to market moving and algorithmic bias's ethical implications are observed in the use of digital twins.

However, the realizations of those actions are set to be defined by the very quick advancing development of data analytics, artificial intelligence, and block chains. These efforts are pursued in cooperation between international institutions, banks, technology companies and other stakeholders trying to establish uniform data formats, standardize security procedures and improve models.

The future of finance and economics seems optimistic, through the prism of digital twins. We can expect to see even more groundbreaking applications in the years to come, such as:

Hyper-personalized financial services: Real time analytics for customization of financial products, investments and insurance policies based on individual needs and risk perception.

Decentralized finance and autonomous trading: Applying the secured blockchain capability to provide personal financial transactions or even automate them through the engagements of the digital twin thoughts basing on the information provided.

Predictive economic forecasting: Through dynamic digital twinning these things allow accurate granular forecasts for the man on the street trends, world happenings and business news.

Building a more sustainable future: By using digital twins, developers create virtual models that provide how green investments, climate change regulations, and resource usage can affect the world; in turn, this assists the globe in advancing to a more sustainable and equitable global economy.

With the evolution of digital twins, those are the ones that need to change. It is in the digital twin that economics shines. Soon we are likely to see even more novel uses.

4.13 Through the Looking Glass: Digital Twins Reimagine Political Systems

Picture looking into the crystal ball that would not only reflect an election night map, but the whole mechanical guts of a political system. An intricate, yet accurate copy, throbbing with information and forecasting the outcome of the elections, the implications of a policy and, why not, of a social unrest. This is the digital twins promise in political systems—a disruptive technology that obscures the line between simulation and reality, which if deeply applied could shift shape to our mode of electing leaders, implementing policies, and democratization. These virtual, but nevertheless very realistic, counterparts to governments, parliaments, and even whole societies are challenging Political Analysis, directing decision-making, and helping to create a more participative and responsive democratic future.

In this environment, the digital child will be a solid analytical model based on sound data replicating the essential aspects of a political system. This can include:

Electoral simulations: The models that include demographics, polls as well as historical voting records offer much more accurate prediction on the next election results.

Policy analysis and impact assessment: Simulations of economies, social landscapes, and environmental systems emulating possible outcome of various policy recommendations.

Social network and sentiment analysis: Online-harvesting of discourse and public opinion—allows you to know how citizens feel and spots social unrest to suppress.

All this data is then inputted into sophisticated AI algorithms and simulation methodologies that end up constructing living, breathing models that can change their behaviour and even forecast possible scenarios. Actors at various levels of the system, whether they be politicians, policymakers, or even normal citizens, therefore can now interact with and study these digital twins to unlock a symphony of underlying dynamics that informs intelligent decision making as they traverse uncertainties about public opinion.

Revolutionizing Politics and Democracy

Digital twins are transforming politics and democracy in profound ways:The first great debate was that of free trade versus protectionism.

Enhanced Policymaking: When policy designers instantiate digital twins of human and environmental systems and simulate the effect of possible policies applicability on different market segment and demographics, more effective and fair policies are designed. The World Bank study revealed that the use of digital twins in policy modelling can increase the growth rate of Gross Domestic Product by as much as 1%.

Greater Citizen Engagement: Citizen participation in policy debates and influencing decisions may be strengthened through virtual town halls, interactive forums, and chatbots enabled by artificial intelligence, thus encouraging a more democratic and comprehensive political system. According to reports by the United Nations e-Government Survey, countries with high participation in e-government are more positively associated with democratic ideals [26].

Improved Crisis Management: Governments use digital twins to create models of possible crises like pandemics, recessional times, national catastrophes, etc., to construct prevention scenarios as well as minimize responses. In fact, according to the research conducted by the MIT Initiative on the Digital Economy, the use of digital twins for disaster preparedness alone may lower economic losses by 30% MIT IDE, 2023, para.

Combatting Misinformation and Disinformation: Digital twins can also be utilized to identify popular internet-driven trends in the social media and consequently identify fake news or propaganda campaigns as a way of mitigating the spread of misinformation hence ensuring an informed citizenry. Another study has revealed that AI-based fact-checking tools diminish the spread of false information via social media by 20%.

Challenges and Opportunities

Despite the immense potential, challenges remain: For Buddhists the ideal is that individuals learn to live in simplicity, detach from material wealth, love the truth, favor peace and justice and renounce violence, killing and war.

Data Bias and Fairness: However, collecting data in a manner, which constructs digital twins as faithful representations of the diverse populations and designing bias-free algorithms, calls for data collection in sensitive methods and ethical considerations.

Transparency and Accountability: The mechanisms of complex political constructions should be open to citizens so that they believe in them and are not manipulated.

Cybersecurity and Privacy: The protection of sensitive political information that's put into digital twins would need effective cybersecurity solutions along with unambiguous regulations on data privacy.

Digital Divide and Accessibility: To avoid this, the digital divide has to be bridged and full inclusivity has to be achieved; thus, all citizens should have equal access to DT phenomenon.

Nonetheless, rapidly growing artificial intelligence, big data analytics, and the development of the blockchain system allow us to solve these problems on a perpetual basis. In view of the government as is technology companies and civil society organizations, the responsible development and use of political system are being prepared for the implementation of the digital twin.

The future of democracy glimmers via the prism of digital twins. We can expect to see even more groundbreaking applications in the years to come.

Hyper-personalized policy platforms: Such platforms should be powered by AI technology and further customize policy proposals and strategies of communication when addressing different segments of voters and their needs.

Decentralized governance and participatory budgeting: Use of the blockchain technology making it possible for the citizens to make decisions that are not only secure but also are free from compromise.

Predictive social sentiment analysis: Active control over public reaction and negative attitude to the policy of the government fulfilling possibility of the opposite influences and conflict resolution.

Adaptive and responsive governance: Digital twins of citizens that are always evolving along with dynamic changes in social and economic worlds and tangent to those dynamic inform the policymakers.

Digital twins might not bring the love of a perfect crystal ball to politics, but they provide a reliable tool for managing the vagaries of the custom bureaucratic timings.

References

1. Ayerbe E, Berecibar M, Clark S, Franco AA, Ruhland J (2021) Digitalization of battery manufacturing: current status, challenges, and opportunities. Adv Energy Mater 12(17). https://doi.org/10.1002/aenm.202102696
2. Ozturk GB (2021) Digital twin research in the AECO-FM industry. J Build Eng 40:102730. https://doi.org/10.1016/j.jobe.2021.102730
3. Wagg D, Worden K, Barthorpe R, Gardner P (2020) Digital twins: state-of-the-art future directions for modelling and simulation in engineering dynam applications. ASCE ASME J Risk Uncertainty Eng Syst Part B Mech Eng. https://doi.org/10.1115/1.4046739
4. Ibrahim O (2019) Digital twin technology: a study of differences from simulation modeling and applicability in improving risk analysis. https://uis.brage.unit.no/uis-xmlui/bitstream/handle/11250/2626223/Ibrahim_Oula.pdf?sequence=1
5. Schrotter G, Hürzeler C (2020) The digital twin of the City of Zurich for urban planning. PFG J Photogramm Remote Sens Geoinf Sci 88(1):99–112. https://doi.org/10.1007/s41064-020-00092-2
6. Chansarkar NA (2020) How digital twins are changing manufacturing. McKinsey, October 29. https://www.mckinsey.com/industries/industrials-and-electronics/our-insights/digital-twins-the-key-to-smart-product-development
7. Accenture (2023) Benefits of digital twins in manufacturing systems, September 27. https://www.accenture.com/us-en/insights/industry-x/manufacturing-systems-architecture
8. Siemens (2022) Digital twins: revolutionizing manufacturing, June 7. https://www.plm.automation.siemens.com/global/en/our-story/glossary/digital-twin/24465

9. Wang W, Yin J, Chai Z, Chen X, Zhao W, Lu J, Sun F, Jia Q, Gao X, Tang B, Hui X, Song H, Xue F, Liu Z, Li J (2022) Big data-assisted digital twins for the smart design and manufacturing of advanced materials: from atoms to products. J Mater Inform. https://doi.org/10.20517/jmi.2021.11
10. PwC (2020) PwC How digital twins can make smart cities better 2020, June 15. https://www.pwc.com/m1/en/publications/documents/how-digital-twins-can-make-smart-cities-better.pdf
11. Tao F, Zhang H, Liu A, Nee AYC (2019) Digital twin in industry: state-of-the-art. IEEE Trans Industr Inform 15(4):2405–2415. https://doi.org/10.1109/TII.2018.2873186
12. Liu X, Jiang D, Tao B, Xiang F, Jiang G, Sun Y, Kong J, Li G (2023) A systematic review of digital twin about physical entities, virtual models, twin data, and applications. Adv Eng Inform 55:101876. https://doi.org/10.1016/j.aei.2023.101876
13. Philips (2023) Healthcare informatics leaders: new care delivery models essential to solving today's challenges and foundation for tomorrow's growth. https://www.philips.com/aw/about/news/archive/standard/news/articles/2023/20230418-healthcare-informatics-leaders-new-care-deliverymodels-essential-to-solving-today-s-challenges-and-foundation-for-tomorrow-s-growth.html
14. Food and Agriculture Organization of the United Nations (2022) IPM and Pesticide Risk Reduction | Food and Agriculture Organization of the United Nations. (n.d.). https://www.fao.org/pest-and-pesticidemanagement/about/understanding-the-context/en/
15. World Meteorological Organization WMO (2023) State of the Global Climate. https://wmo.int/publicationseries/state-of-global-climate-2023
16. Rocky Mountain Institute (2023) Annual report 2023 – RMI. RMI. https://rmi.org/annual-report-2023/
17. Renewable Energy Statistics – IRENA (2022). https://www.irena.org/publications/2022/Jul/Renewable-Energy-Statistics-2022
18. McKinsey & Company (2023) Betley, B., Cariati, T., Gao, F., Hannon, E., Levy, C., Parente, F., & Seo, J. Cloud-powered technologies for sustainability. https://www.mckinsey.com/capabilities/mckinsey-digital/ourinsights/cloud-powered-technologies-for-sustainability
19. US Army (2023) US Army moves out on digital engineering strategy. Defense News. https://www.defensenews.com/land/2024/06/19/us-army-moves-out-on-digital-engineering-strategy/
20. RAND Corporation (2022) Using technology to improve the agility of force generation processes. In Frank AB, Bartels EM (eds) Adaptive engagement for undergoverned spaces: concepts, challenges, and prospects for new approaches (RR-A1275-1, pp. 397–398). RAND Corporation
21. NATO (2023). https://www.nato.int/cps/en/natohq/official_texts_217320.html
22. UNEP (2022) Annual Report- UN Environment Programme. https://www.unep.org/resources/annual-report-2022
23. IOC (2023) The potential of digital twinning to drive Games optimisation. Olympics.com. https://olympics.com/ioc/news/the-potential-of-digital-twinning-to-drive-games-optimisation
24. Stainley (2022). https://www.morganstanley.com/content/dam/msdotcom/en/assets/pdfs/Morgan_Stanley_2022_ESG_Report.pdf
25. Mastercard (2023) Mastercard expands first-of-its-kind AI technology to help banks protect more consumers from scams in real time. Mastercard Newsroom
26. DESA, United Nations (2022) E-Government Survey 2022. https://desapublications.un.org/sites/default/files/publications/2022-09/Web%20version%20EGovernment%202022.pdf

Chapter 5
Modeling of Digital Twins

5.1 Introduction

Many researchers have focused their efforts on developing and expanding upon the concept of digital twins since the concept's inception, paying significant attention to the architectural aspects of the concept. In a four-dimensional Digital Twin architecture that includes models for Virtual Data collection and transmission, Estimation, Control and Virtual Modelling. It lacks the view of the interaction between the user-model and the overall process of user-model interaction and hence does not include the concrete global view of the modules. However, these frameworks cannot encompass all the subtleties of digital twin model, but developers have an opportunity to define a conceptual model of twin through the reassembly of the modules. For instance, the three-dimensional model does not contain a field tagged as 'Environment' so they are not very helpful in this case [1, 2].

However, a five-dimensional framework to enrich the current digital twin as well as the relationship between both halves: the physical and the digital one. It determines the ability of the user to interact with each of the modules and their interaction with the other modules within the digital twin model while outlining the working model of the model. One more essential paradigm of the modelling process is the structuring of avatars and applications concerning real and virtual objects and objects relationships [3] (Fig. 5.1).

The digital twin creates a linkage between the virtual and real worlds since it serves as a representation of the two. However, the methods of digital twin modeling at the moment remain insufficient in their capacity to offer widespread support, primarily because it's modeling heavily employs modular representation. These methods cannot be used in designers' fields of work as they do not provide enough information on the analysis of components of the digital twin and its specifics. It is, therefore, important to proceed methodically to develop a conceptual-level model of the multi-dimensional digital twin since, as pointed out earlier, digital twins are

© The Author(s), under exclusive license to Springer Nature Switzerland AG 2025

S. Gupta et al., *Digital Twins*, https://doi.org/10.1007/978-3-031-76564-3_5

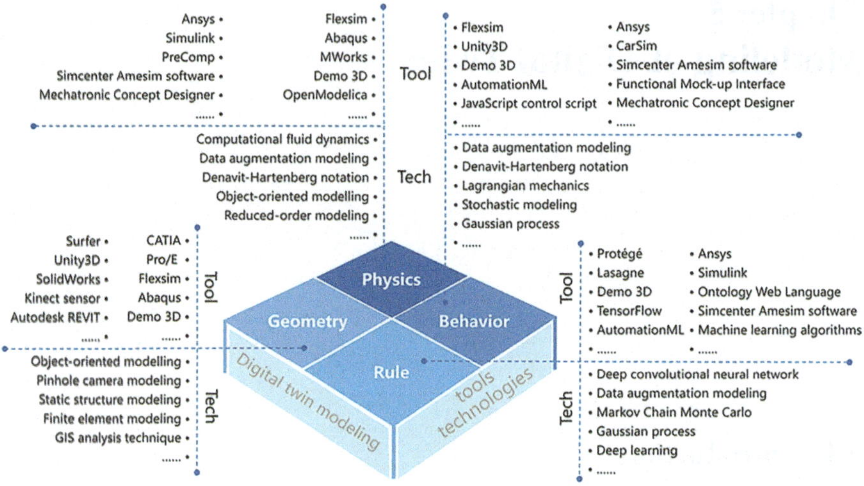

Fig. 5.1 Technological framework and toolset for constructing Digital Twin models. (Source: [1])

increasingly being employed in engineering disciplines. The present chapter aims to show stepwise how a conceptual model of a digital twin is developed [4].

5.2 Conceptual Design Development

Building an effective digital twin is based on a coherent concept of the idea. Purpose & Scope to start with the digital twin. Define the thing being modeled—a specific physical asset or process, identify functional requirements e.g., predictive model describes level of fidelity and accuracy that needs to be maintained in the balance of computational efficiency.

Selection, Data Acquisition, Management and Design, these include sourcing data to be used, typically via sensors or from historical records, determining a strategy for data collection, and implementing pre-processing steps to confirm the data quality.

Modeling and Simulation, the selection of modeling technique is dependent on the complexity and fidelity you want to achieve. On the one hand, physics-based models have the potential to achieve high accuracy, but can be computationally expensive, particularly for larger-scale, complex systems; on the other hand, data-driven models are generally less expensive to develop but are often less interpretable. It must be able to scale in the future and include practices for model updates and data integration to keep the digital twin alive and accurate [5].

Accordingly, based on the five-dimensional framework, the conceptual digital twin model of electric vehicle is having following specific steps [6, 7]:

Step 1: Function analysis. While using this intelligent vehicle, inputs concerning the pedestrians and other vehicles in the environment were fed into the vehicle. Power was needed for normal working. Since safety was an issue, path planning which was enabled through data analysis was needed. Consequently, the driving decision was the terminal product. The overall objective of the vehicle was the formation of self-driving behaviors.

Therefore, in the case of efficient path planning, that is, the intelligent vehicle, all of its functions must cooperate. It required positioning of the vehicle, recognition of pedestrians, and obstacles, recognition of traffic signs, behavior decision model. Thus, the total function

of the vehicle was analyzed into location function, obstacle detection function, pedestrian detection function, traffic sign identification function, and decision behavior.

Step 2: This means that the functions shall be divided and placed in a number of different domains. As for the location decision, the positional information of the subject was identified without any directions from the vehicle. This was a function of as a measurement of the actualization of the entity's potential for excellence. The following, namely obstacle detection, pedestrian detection, and traffic sign identification meant that road conditions needed to be observed and data surrounding the issue acquired. Concerning the detection of obstacles, the vehicle required communication with the driver to reduce risks. Thus, obstacle detection is a function of system security. Decision-making incorporates the stage of assessing the available data and formulating relevant instructions, and that makes it fit into the classification. When it comes to leading indicators, from our point of view, it appears that data are engaged in such functional activities as location decisions, obstacles' identification, detection of pedestrians, and traffic signs. These functions should also fit under the well-established label known as the Decision Diary or the DD.

Step 3: Due to the function analysis, the components can be described based on each of the functions they perform. For instance, the battery served as energy source for the car, the positioning function was fulfilled by GPS and inertial navigation system [7] (Fig. 5.2).

5.3 System Modeling and Understanding

Digital twins, virtual replicas of physical entities, have revolutionized fields ranging from manufacturing to healthcare. Central to their efficacy are mathematical models, which underpin the simulation, analysis, and prediction capabilities of digital twins. The section focusses on the indispensability of mathematical models to digital twin simulations, emphasizing model building, application, and implications.

Definition: A digital twin is an evolving digital replica of a living or non-living physical entity which relies on machine learning, sensor calibration and data analysis to allow it to operate, monitor and manage the device or system connected to it.

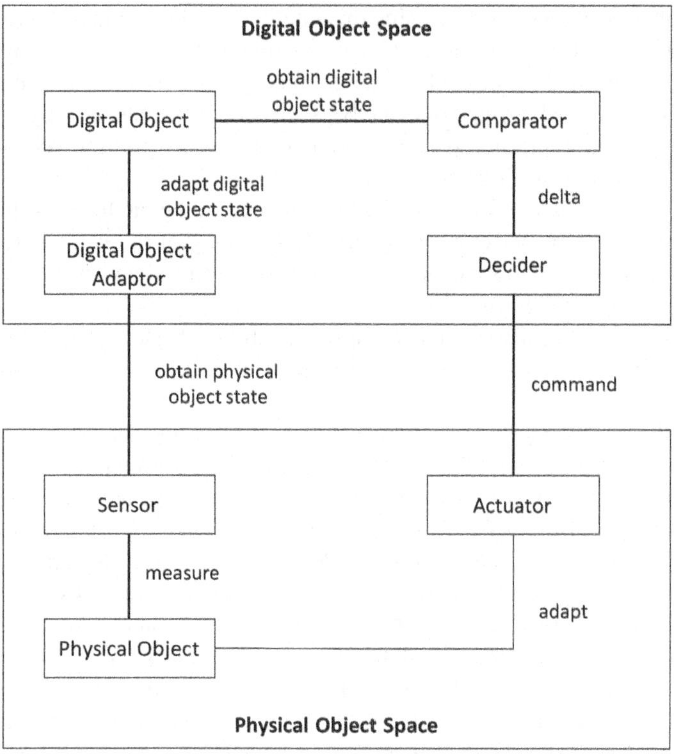

Fig. 5.2 Proposed conceptual modeling of digital twin. (Source: [6])

At the heart of the idea is a continuous dialog between the physical and the virtual world, to guarantee an accurate representation in the digital twin of the physical counterpart. Models created by means of mathematical concepts and language to reflect the system under study They range from simple models, such as a series of equations, to more complex models made up of numerous algorithms that describe all types of physical processes. Digital twins rely on them as they simulate the physical behaviour and routines of the real world [8].

Understanding complex systems is a fundamental problem in many branches of science and engineering. From the subtle workings of a cell to the global scale of the climate, physical systems can display highly complex relationships and behaviors. In essence, physical systems are composed of interacting components that have cause-and effect relations between them. Yet, directly observing and analysing these interactions can be cumbersome, particularly in large-scale or highly dynamic systems. This is where mathematical models are useful. Mathematical model: a great simplification of the mathematical abstract representation of a physical system. This abstraction involves identifying the key components and relationships within the system, and then translating them into mathematical equations, functions, or other formalisms. The advantage of this approach lies in its ability to capture the essence of the system's behavior while filtering out irrelevant details [9].

A Spectrum of Techniques

On top of these general kinds, the idea of the digital twin includes myriad specific methods and mathematical modeling techniques: For example, Ordinary Differential Equations (ODEs): describes how a system behaves over time (useful for things like fluid flow or heat transfer, i.e. continuous processes). Partial Differential Equation (PDEs): It is useful in dealing with systems that evolve over space. They are aimed at simulating how the properties of a system vary over space and time and demonstrate applications that range from the simulation of something like stress distribution in a bridge to that of wave propagation in an aircraft wing.

Agent-Based Models are appropriate for systems with interacting entities (e.g., traffic flow, manufacturing production lines) Specifically, these predict how each entity will behave both as an individual and in terms of how they will act in the environment and with respect to other entities. Machine learning algorithms are used to extract patterns and relationships through sensor data. These patterns are then ripe to be twisted in any wishful data-driven or physics-informed models leading to their higher-accuracy and generalization.

The choice of the modeling technique entirely depends on the application and the fidelity expected. Hybrid one of the widely types of research which going on this tip using both way for prediction of traits which are having hybrid techniques based on physics and data mining, by this way one gets the efficient, accurate and robust approach to do the task. Finally, emulating complex system, requiring excessive computational load for real-time implementations, its model downsizing is required [10, 11].

Essentially, the most obvious manner of creating physics-based models is in creating a set of equations that directly correspond to the governing laws in physics (Newton's Laws, Maxwell's Equations, etc). With an emphasis on being able to generate idealized simulations and a high cost computationally as a result, it comes at the price of difficulty in implementing, especially when applied towards more sophisticated systems.

Data Driven Models These models understand the relationships in the system statistically since they work on the data collected from the sensors and various other sources. While this may be cheaper to train and deploy initially, these models are unlikely to capture the governing physical principles and therefore, these models are unlikely to generalize and are not particularly strong at predicting future cases. So, in the example of jet engines, the model would be trained on the sensor data of multiple points within the engine when it will be in operation. These could include values such as temperature, pressure, vibration, or fuel flow at a variety of engine speeds and/or altitudes. These would then be fed into machine learning algorithms like decision trees or neural networks in order to detect patterns in this historical data with respect to the relationships between these sensor readings and engine performance metrics such as thrust as well as fuel efficiency. There is no question that the benefits of being data driven are clear. Although firstly, data-driven models are sometimes computationally cheap and faster to develop than the physics-based models. Equally as important, they are designed to quickly adjust to new sensor data that improves the fidelity of the model as new sensors add perspective to how the engine is behaving or operating. This could be the ability for a model to recognize the small noise patterns of a failing bearing to ultimately predict engine train availability and save the cost of catastrophic engine failures.

But data-driven models have their shortcomings. This is usually because they are not very good at explaining the "why" regarding some engine behavior. They may find increments but provide little explanation about the underlying theory behind this. In addition to this, the skills of these models are dependent on the quality and the quantity of the data that they have gotten. Incomplete Data or Bad Quality Sensor Readings—can lead to flaky models Furthermore, the behavior of models trained on particular operating regimes for example take-off at sea-level might exhibit poor performance when the engine is in different regimes altogether for example flying at high altitude [12].

There are multiple benefits of hybrid modeling. It mixes the high-fidelity of physics-based models with the flexibility of data-driven models to enhance the precision. In addition, the computational cost is typically much less than a purely physics-based model. Additionally, the fact they are physics-based should give the model good performance on new scenarios, whereas the data-driven aspects will give some explainability by visualizing the effect of the sensor readings on the engine performance.

In addition to jet engines, hybrid modeling will be applicable to other sectors such as the following categories. Hybrid model- This can be used in Predictive Maintenance in Manufacturing that combines physics-based information on the wear and tear of the machines with data from the sensors to predict the possible failures and thus make maintenance schedules for machinery at factories more optimized [13].

Models define the degradation based on certain inputs or data. In place of the physically based models the data-based models use data. To identify dynamical systems from sensor data measured in a physical system, three model-based data algorithms are designed.

In fact, surrogate modelling and system identification are more alike than one might think. The purpose of system identification is to minimize certain forecast errors regarding the characteristics of a model, particularly within its class. It has to be understood that the goals of a surrogate model are to minimize the forecast errors and, at the same time, adequately model the behavior of the actual system. System identification employs real-time data acquired from sensors as well as experiments to adjust parameters aiming at unveiling a solution, in contrast to surrogate modelling which is developed and gradually updated employing the outcomes of various simulations conducted at different times. Due to the similarities, the presented system identification techniques/strategies and surrogate modelling methodologies can be used interchangeably.

5.4 Digital Twins by First Principle Modeling

A digital twin is a virtual replica of a physical entity that uses real-time data to simulate the behavior and performance of its real counterpart. They are expected to have a wide range of uses, from helping manufacturers better manage product lifecycles, to helping assess when infrastructure items may require service. First-principle based modeling is a fundamental methodology to build a digital twin. These include laws and relations that are the foundation for a system, known as first principles. First principles modeling is the construction of a mathematical model of the physical system from these fundamental principles for the digital twins. This includes aspects like, Physics—To model the real physical behavior of the system that follows the laws of motion, thermodynamics, electromagnetism etc. Chemistry—Inclusion of material characteristics, chemical reactions, etc., used to model material behavior. Engineering perspectives are modelled using stress-strain relationships, fluid dynamics, etc. [14].

5.4.1 Pros of First Principle Modeling

High Fidelity: Models in this category model the underlying physics of the system and have the potential to achieve a high degree of accuracy. This makes simulations very realistic and allows for emulation of the output.

Explanatory Power: As these models are constructed based on the physical principles, they help in understanding the roots of why the system behaves in the way it does. This provides a more comprehensive view of cause-effect relations.

Prediction: First principle models allow ex-ante simulation of a range of different scenarios, for example, how the physical system will behave if we vary certain input conditions. This will also help in proactive decision-making.

5.4.2 First Principle Modeling Difficulties

Developing first principle models are always quite complex and requires strong expertise in the underlying physics and engineering principles. Simulating these highly detailed models can become computationally expensive, and this is especially the case with complex systems.

Real-world systems involve uncertainties and for modeling purposes, some simplifications must be made due to model uncertainty.

Aerospace Engineering- Aerodynamics and other first principle models are used to model the flight dynamics of aircraft. Digital twins of power grids can be developed from first principles that represent the power flow of electricity, alert system beforehand to prevent a possible electrical failure, and optimize the operation of the grid to achieve service continuity. Chemical Engineering-first-principle models are used to simulate chemical reactions in reactors and optimize process efficiency and product quality. First principle modelling is a very high-fidelity way to create digital twins however this is undesirable for non-physically significant aspects of the model. Despite its limitations in complexity and computational cost, the interpretability and ability to understand and predict using the data have powerful applications across industry combines. With more computational resources and evolving modeling techniques, the first principle modeling is only set to play a more significant part in the future of digital twins.

5.5 Mathematical Models Form the Basis of Many Digital Twin Simulations

A mathematical model can be precisely described by an expression, formal language, or a variety of expressions that describe the behavior of an ideal being. To the software developer within the context of a digital twin, these models are that part of the physical laws, engineering rules, and operative behavior that an actual individual implemented. Mathematical models range in terms of scale and the two primary types of models are used in digital twins [15]:

The power of mathematical models lies in several key aspects. Firstly, they provide a quantitative framework for understanding system behavior. By expressing relationships through equations, models allow us to predict how the system will respond to changes in its inputs. For instance, it enables us to make experiments and simulations for the studying of the dynamic behavior of the physical system, such beneficial aspect being produced by the fact that there is no necessity for the interference with the physical world since it could be expensive, time-consuming or, in some cases, even lethal. May also help to analyze and reason about the system (e. g., mathematical). Using a certain mathematical tuning, primarily, basic calculus-differential equations, how these equilibrium positions and how equilibrium is attained can be explained for the understanding as to whether the given system in question is stable or sensitive or is always in equilibrium. Thus, it becomes possible for scientists to understand what may be needed for system functioning, or what can be wrong with a particular system, as well as to deal with a system performance as efficiently as possible.

What mathematical model applies here depends of the nature of the system under consideration and the level of resolution. In basic models, system behavior patterns are emulated across the world; on the other hand, in detailed models, exact emulation is envisaged. Therefore, there will always be a trade- off between models' accuracy and the ability to perform computations at any level of hierarchy. If ever there is a need for a system of mathematics to understand a given model, then that model is actually of no use but if a model is very simple then a lot of facets of the actual system cannot be captured.

Mathematical models are used in the physical sciences in a more profound way in the modeling and understanding of systems in the real world. The utility of models is that models are the way of logical and numerical analysis of the complex systems and with the help of these models, one gets a quantitative idea about how such systems are being managed, a level of understanding which can not be attained by observation or experimentation. The world we will describe in terms of the complex and I think that an interesting world which we are into. The companies are as if they gain size in proportion with their size it seems to increase.

Challenges and Opportunities of Mathematical Model
Mathematical Modeling for Digital Twins systems at work in the Real World are nondeterministic. This provides a model that accounts for these uncertainties to make sound predictions. Optimal solutions to scaling out and normalizing the bindings so that the models can be natively integrated on large, federated systems. Simpler models will make it easier to explain why the model has taken the actions it has and how much we can still trust the model for human decision making. Regardless, the future of mathematical modeling of digital twins is bright. Such fast advancements of digital twin innovation were possible not only due to improved computational power, but also with the help of AI and machine learning techniques for significantly more accurate and thus validatable models and standardized digit twin approaches that make digital twins per se feasible, meaningful, and user-friendly [16].

5.5.1 Development of Mathematical Models for Digital Twins: Model Formulation

Based on the analyzed data, mathematical models are formulated. This process involves defining the variables, parameters, and equations that represent the system's behavior. Various modeling techniques, such as differential equations, finite element analysis, and agent-based modeling, can be employed depending on the system's complexity. The creation of a digital twin begins with the precise specification of a mathematical model that offers a well-documented and often quantitative representation of a physical system by listing variables, parameters, and their affecting equations representing and explaining the properties and characteristics of the actual system. Through these stages, systematics can be expressed with greater precision and in greater detail to create accurate digital models of products in production or heat transfer, etc. These will enable digital twins to forecast the future statuses and possible issues and improvements just by extrapolation of current and prior state data. The ideas from the digital twins are applied to determine the relations between elements of the system of system like healthcare, where patient's digital twins might explain existence of physiological community for disease identification and treatment. Moreover, sensors are incorporated in the mathematical models for capturing and interpreting the raw data for observing the current state or modeling the nearest future states and applying process enhancement, as done in smart manufacturing. These models also support scenario analysis and optimization, which if discerned by urban planners are important to determine which strategy is better for certain situations such as traffic density, energy usage, pollution, etc. The next level of specificity is quantitative, and that is where mathematical models come in and it gives an industry like Oil and Gas another level of specificity when our system is considered at large and this will help in developing strategies. Thus, mathematical model formulation ensures digital twins are powerful tools for simulation, analysis, and optimization across various sectors.

Providing specific mathematical equations for each example would require a detailed examination of the respective digital twin implementations. For precise physical modelling in a digital twin, data-driven models are essential, particularly in two key cases: This occurs when physical processes demand a strong model which is either too robust or not well-enough understood or when the actual structures of the connected items are complex and difficult to define. Physics can be explained via physics-based modelling which will require a lot of CPU time and resources while it can be simulated through the software available today. Another issue is that the digital twin might need complex iterations of computations that the physics model needs. To capture the essence of the primary objective it is necessary to develop elementary models that represent the input/output relation accurately. Consequently, the predictions on the digital state can be optimized and managed in real-time using a digital twin.

5.5.2 Statistical Models

For inference and prediction Statistical and ML models have been widely used in the present era. Whereas ML seeks for the prediction through the employment of generic learning algorithms to disparate but massive and sometimes complex data sets, the statistical methods as their methodology of work have highlighted inference as their primary goal and can be described as a process of fitting a certain probability model to data as it occurred in the framework of a specific project. In this part let us explain some of trends that continue to shape the field of data-driven modeling and investigate some of the examples of how some of these trends have been implemented in the selected digital twin examples. The crucial factor concerning the building of connections between P2V (Physical to Virtual) and V2P (Virtual to Physical) in terms of digitized twins is the statistical explanation. Using the results of the dynamic system identification, the state space models are proposed with regard to measurement data on P2V connection. Diagnostic and prognostic of failure and time-based maintenance of physical systems based on statistical models for the V2P link and modeling degradation.

Probability-based approaches to converting physical systems into dynamic systems (P2V). The method of determining a mathematical model for a dynamic system using two measurable input and output data is known as system identification.

As systemic engineering progressed in the last decade, the reliance has been more inclined toward the employ of statistical methods to system identification particularly for linear dynamic systems [17].

Further, to differentiate between the 'chaotic' systems for which the behavior will be restricted and random from other systems the researchers have tried simple time domain ARMA models. For the other non-stationary dynamic systems, we have employed auto-regressive integrated moving average (ARIMA) models for non-stationary cases beyond ARMA method complications of the stationary case only. The ARIMA model is more of an elaborative statistical model compared to Autoregressive, Integro, Moving Average because the trending data can be represented in the form of a sequence of difference. There is a few other ways of performing time-domain system identification and they include the following: the eigensystem realization algorithm and several forms of stochastic subspace system identification (SSI). Among the frequency domain models, some of the most frequently used method is FDD (Frequency Domain Decomposition) technique. SS is a state-space identification method, and this method employs stochastic models for state-space models for identification of systems; SSI has remained greatly popular out of all the mentioned linear, system, time-domain, and frequency-domain methods. Over the last decade, various changes to the SSI algorithm have been made that helped create other new SSI's including the reference-based SSI and the covariance driven SSI [18].

5.6 Data Driven Modelling [10, 12]

Data-driven modeling provides a powerful means to develop digital twins, especially when the physics of the real-world system is poorly understood in great detail or when computational efficiency is critical. Unlike physics-based models that require detailed equations that replicate every law of physics in play; data-driven models exploit years of sensor data, logs, and the likes. A data-centric approach has many benefits. For example, the amount of calculations these two data-driven based models require are typically lower, making them excellent for application in real-time applications, where computational power may be more limited. Secondly, physics-based models generally take longer to develop relative to machine learning, particularly when a comprehensive understanding of the underlying science is not as mature. The biggest advantage of data-driven models is that they are by definition flexible. The model is an ongoing work in progress that is constantly updated with new data and as such that is becoming optimized more and more towards the reality during the time of the model. But realizing the true power of data-driven modeling isn't as simple as it sounds, a process must be carefully defined. This is followed by an essential data acquisition exercise in which all historical sensor readings, data related to the operation, and any other agents that control the way the system behaves are collected. The aforesaid data then goes through pre-processing to make it neat and clean, the pre-processing involves handling of missing values, outlier, inconsistency etc.

The major requirement for any data-driven work is to have data ready to work on, followed is the selection of an appropriate data-driven technique. Machine learning algorithms such as decision trees, and neural networks, are also able to find the complex structures in the data which may later help to produce predictions for the future behavior of the system (predictive maintenance, optimizing the operations), find anomalies. Statistical modeling techniques such as regression analysis add yet another complement to the model by establishing connections between the variables. Time Series Analysis—It looks at the data collected through time and helps in forecasting the future values and detecting the trends that can be used for proactive maintenance, and resource management. But when these methods are combined with one another—as well as updated with the current date consistently—they come together to form an increasingly accurate model with new data; a data-driven model twin which can provide insights into the real-world system that it represents. Use cases include predictive maintenance in industrial environments, energy consumption optimization in smart buildings, and logistics and supply chain management optimization. All of this is nothing of all things but how unaccompanied to admit the constraint of information driven modeling. Whether the models are accurate or not is relying particularly on the quality and the amount of the data collected. Inevitably, poor quality data will translate to poor quality, unreliable models. Moreover, they cannot offer explanations on why a system exhibits a certain behaviour but are rather designed to simply identify existing correlations rather than fully understanding the underlying physical principles.

The models trained in one set of operational conditions may not extrapolate to other operating conditions. This in mind, data-driven modeling is a tempting platform for the creation of digital twins and by the process too, the efficiency and versatility of this method of digital twin creation makes it a valuable asset in various industries. With ongoing improvements in both data collection and analysis procedures, it appears that data-driven modeling can soon play a pivotal role in further exploiting the potential of digital twins and helping us make more sense of complex systems in the world.

Regression Analysis This method creates a model to connect the dependent variable (e.g., the engine performance metric) with one or more independent variables (e.g., multiple sensor readings) i.e. T (temperature), P (pressure), V (vibration) and F (fuel flow) This can be represented as:

$$Y = \beta_0 + \beta_1 * T + \beta_2 * P + \beta_3 * V_3 + \beta_4{}^* F + \varepsilon,$$

$$Y = \beta_0 + \beta_1 * X_1 + \beta_2 * X_2 + \beta_3 * X_3 + \beta_4 * X_4 + \varepsilon,$$

where Y is the engine performance metric and the coefficients β_0, β_1 to β_4 are the regression coefficients for each sensor data point, and ε is the error term. The model can interpolate or extrapolate engine performance based on real-time sensor readings by solving for the regression coefficients.

Algorithmic Integration

Machine learning algorithms and statistical models create digital twin templates by training on (rich, nebulous) datasets. The models consume real-time sensor data from the engine. The models will process the data and predict future breakdowns, engine diagnostics, or any other abnormal pattern. Engineers, or pilots, see these forecasts, and plan and decide accordingly.

Safety and Anomaly Detection: Historical data can be then used to statistically define the operating range of normal for each individual reading of each sensor. These ranges are used to compare range to new data points from the sensors. Most relevant alerts can be generated if there are major deviations, which can also result in potential safety issues.

Benefits (Mathematically)

Predicting failures means spending less time towing: Downtime Cost = Cost per Flight Hour * Number of Downtime Events (reduced by prediction).

Fuel Cost = Fuel Consumption * Fuel Price (fuel consumption is a mathematical model based on sensor readings; optimizations make the engine consume less fuel).

Digital twins that use data, algorithms, and math to derive insights; data-driven digital twins through continual updates and feedback, the digital twin is able to drive exceptional advancements in efficiency, safety, and cost reduction in the aerospace sector. The capabilities of these algorithms will only continue to expand in scope and reliability with advancing data collection and analysis methods, this predictive technology will soon potentially be even more transformative as we head into the future.

The idea of a data-driven modeling inside the digital twins is not something only of the future but is an evolution that would help completely reshape the way we work with and maintain the complex systems like an aircraft engine. With advancements in data collection and analysis only just beginning to permeate so many industries around the world, the future of aerospace offers more incredible developments fueled by the benefit of digital twins and the understanding that they bring.

A Data-Driven Approach to the Aerospace Digital Twin

In a world where we could see airplanes not only fly but their digital twins as well fly right alongside them. The key to achieving maximum engine performance, flagging potential failures, and optimizing maintenance schedules lies in the virtual twins that learn and evolve continuously. It's exactly how the aerospace industry stands to benefit from data-driven modeling in digital twins.

From Sensor Data to Autonomous Insights: This approach leverages the rich sensor data available from modern aircraft engines. At any given time, the engine is monitored for parameters such as temperature, pressure, vibration, and fuel flow, providing a play-by-play of the engine's health. This data strait is the fodder to transmission models in a digital twin.

Machine learning is one of them and is a very potent technique in use. These algorithms such as decision trees and random forests are trained on historical data from sensors to learn patterns and correlations between different parameters. These patterns are then used for predictive maintenance. The digital twin will analyze the real-time sensor readings and compare them with the learned patterns to predict when something may fail soon if not addressed beforehand. Now, picture the digital twin sounding the alarm years in advance of a key part of failing, attempting a no-thrills repair rather than a spectacular in-flight breakdown.

They are used in concert with traditional statistical modeling techniques such as regression analysis. The digital twin is able to predict failures in addition to being able to monitor day-to-day engine performance—a power that arises by correlating certain sensor readings with specific engine performance metrics. For example, the model may identify a series of sensor readings that suggests suboptimal fuel burn. This data can help identify trends that engineers can use to make small changes to engine settings or routing directions to reduce fuel costs.

One of the major advantages of using data-driven modeling is that it is, by its very nature, highly adaptable. Each successive flight provides updated data, which in turn teaches the digital twin to refine its models. This helps ensure that the digital twin remains useful and representative of how the engine is performing in the real-world over its lifetime and in varying conditions.

Predicting anomalies is not only limited to maintenance issues. A more semantic approach is to train models on sensor readings, which will help emphasize a reading that is far from the normal range. So, it could represent a possible safety risk, allowing the aircrew and ground staff to intervene. The ability to detect these anomalies ahead of time can prevent disaster and protect passengers and crew.

5.7 Mathematical and Algorithmic Perspective on Data-Driven Digital Twins in Aerospace

Data-driven digital twins play a critical role in enhancing the performance of aerospace applications by using sensor data, and mathematical & algorithmic models to improve the fidelity of digital twins. A more in-depth look into this process is found here.

Input Data: The Dataset for Insider Threat Recognition (DIRNDL) is used, which is used to simulate insider threats at fighter jet sites. Engine parameters such as T (temperature), P (pressure), V (vibration), F (fuel flow) (sensor data captured at high frequencies). The structure of this data would be a time series matrix rather than the traditional data matrix [19–21].

$$X = [[T_1, P_1, V_1, F_1],$$

$$[T_2, P_2, V_2, F_2],$$

$$\ldots$$

$$[T_n, P_n, V_n, F_n]]$$

where n is the number of data points gathered.

Preprocessing: Missing values are filled, outliers are treated, and data is normalized so that all the features have the same scale.

Machine Learning Algorithms

Decision Trees: These are the set of algorithms which form a tree like structure and each node in the tree represents a split on a sensor value. This tree is constructed by recursively splitting the data into octants of this hypercube, with a split ('node') at each junction, until each leaf contains only one data point (reading). The model learns how to classify new engine readings by navigating the tree via these splits. This allows for the detection of patterns for different sensor readings known to be present in the context of anomalies or suspected failures.

Random Forests: It performs the same task as Decision Trees; however, instead of creating one single Decision Tree with the entire dataset, it creates an ensemble (set) with a ton of them, where it takes a part of the predicted result of all of them to output the final prediction ratio.

Combining the predictions from all trees makes final predictions more accurate and stable than a single decision tree.

Sequential Learning Algorithms

Incremental Updating: Online Learning Similar to the stochastic gradient descent method to dynamically update machine learning models. Whenever new data points (sensor readings from each flight) are available, the model parameters are updated such that the error between the model's forecast and the true engine behavior is minimized.

Random Forest

The Random Forest algorithm consists of an ensemble of decision trees. Each tree is built independently based on a random subset of the training data. The final prediction is obtained by aggregating the predictions of individual trees.

Mathematically, let **X** be the feature matrix representing historical sensor data, and **Y** be thc target variable indicating the occurrence of a failure mode. Each decision tree T_i is built recursively by selecting the best split at each node based on a random subset of features. The prediction $\hat{Y}$ for a new observation **x** is obtained by averaging the predictions of all trees in the forest:

$$\hat{Y} = \frac{1}{N}\sum_{i=1}^{N} T_i(\mathbf{x})$$

Gradient Boosting Machines (GBM)

Gradient Boosting Machines sequentially build an ensemble of weak predictive models, typically decision trees. Each new model corrects the errors of its predecessor by fitting to the residuals.

Mathematically, let $F_m(x)$ be the ensemble of m weak learners. The prediction of the final model is obtained by summing the predictions of all weak learners:

$$\hat{Y} = \sum_{m=1}^{M} F_m(x)$$

Neural Networks

Neural networks are a form of connectionist systems that are characterized by interconnected layers of neurons that perform computations on the input data to offer output predictions. Algorithms like feed forward neural net or Recurrent Neural Networks (RNN) are capable of learning from raw sensor data or from specific patterns. Mathematically, the forward pass of a neural network can be represented as:

$$\mathbf{z} = \sigma(\mathbf{W}_1\mathbf{x} + \mathbf{b}_1)$$

$$\hat{Y} = \sigma(\mathbf{W}_2\mathbf{z} + \mathbf{b}_2)$$

Where:

- $\mathbf{x}$ is the input data.
- $\mathbf{W}_1$ and $\mathbf{W}_2$ are weight matrices.
- $\mathbf{b}_1$ and $\mathbf{b}_2$ are bias vectors.
- σ represents the activation function.

5.7.1 *Statistical Models for Correlating Operating Conditions with Maintenance Requirements*

Regression Analysis

Linear regression models quantify the relationship between independent variables X (e.g., operating conditions) and a dependent variable Y (e.g., maintenance requirements) using a linear equation:

$$Y = \beta_0 + \beta_1 X_1 + \beta_2 X_2 + \ldots + \beta_n X_n + \epsilon$$

Where:

- $\beta_0, \beta_1, \ldots, \beta_n$ are the regression coefficients.
- ϵ is the error term.

Survival Analysis

The survival analysis models handle the time related data that is time to event, for example, time to failure or time to maintenance. The Cox proportional hazards model is one of the most popular and commonly applied methods of survival analysis. It models the hazard function $\lambda(t)$ as a function of covariates X:

$$\lambda(t|X) = \lambda_0(t)\exp(\beta_1 X_1 + \beta_2 X_2 + \ldots + \beta_n X_n)$$

Where:

- $\lambda_0(t)$ is the baseline hazard function.
- $\beta_1, \beta_2, \ldots, \beta_n$ are the regression coefficients.

Time Series Analysis
Time series models work with series data collected for the analysis of temporal characteristics of sensors and maintenance records. Among the existing time series models which can be used for forecasting the requirements for maintenance involving historical trends and seasonality, the most popular is the ARIMA model.

$$Y_t = \beta_0 + \beta_1 Y_{t-1} + \beta_2 Y_{t-2} + \ldots + \epsilon_t$$

Where:

- Y_t is the value of the time series at time t.
- ϵ_t is the error term.

These mathematical formulations help in forming a basic foundation in order to allow machine learning algorithms or statistical models used in predictive maintenance systems which gives organization a capacity to predict failure modes and to define the relationships between operating conditions and necessary maintenance needs rather well.

(i) *Healthcare: Personalized Medicine—Biomechanical Models:*

Biomechanical models are used to construct CFD models focused on reconstructing and analyzing blood flow within the heart. Equations often used to describe blood flow rate in the heart [20]:

5.7.2 Computational Fluid Dynamics (CFD) Equations for Blood Flow in the Heart

Continuity Equation
In an incompressible fluid such as blood the continuity equation is analogous to the conservation of mass meaning when it comes to blood flow in the heart it would be a statement that the rate at which blood is entering a location is equal to the rate of blood that leaves that location.

$$\nabla \cdot \mathbf{v} = 0$$

Where:

- $\nabla \cdot$ denotes the divergence operator.
- $\mathbf{v}$ is the velocity vector field representing blood flow.

(ii) *Navier-Stokes Equations:*

The Navier-Stokes equations describe the conservation of momentum for fluid flow. In the context of blood flow in the heart, these equations account for the effects of viscosity, pressure, and inertial forces on the flow patterns.

Momentum Equation (Conservation of Linear Momentum)

$$\rho\left(\frac{\partial \mathbf{v}}{\partial t}+(\mathbf{v}\cdot\nabla)\mathbf{v}\right)=-\nabla p+\mu\nabla^2\mathbf{v}+\mathbf{f}$$

Where:

- ρ is the density of blood.
- $\mathbf{v}$ is the velocity vector field.
- t is time.
- p is the pressure.
- μ is the dynamic viscosity of blood.
- $\nabla^2\mathbf{v}$ is the Laplacian of the velocity field.
- $\mathbf{f}$ represents any external body forces, such as gravity.

(iii) *Constitutive Equations:*

Constitutive equations relate the stress tensor to the strain tensor, characterizing the mechanical behavior of blood. These equations are often simplified for blood flow simulations, assuming a Newtonian fluid behavior due to the low shear rates encountered in blood flow.

Newtonian Fluid Model

$$\tau = -\mu\nabla\mathbf{v}$$

Where:

- τ is the shear stress tensor.
- μ is the dynamic viscosity of blood.

These equations, along with appropriate boundary conditions and geometry descriptions, form the basis of computational fluid dynamics simulations of blood flow in the heart. By solving these equations numerically using techniques such as finite volume or finite element methods, researchers can simulate and analyze various aspects of cardiac hemodynamics, aiding in the understanding of cardiovascular diseases and the design of medical interventions.

5.7.3 ML (Machine Learning) Models: ML with a Physics Background [22]

It is possible for first-principal simulation data to coexist with physical systems data in some of the physics-related components of DTs. They provide the basis for innovating the approaches to modeling such complex physical systems with the help of the ML. For instance, the first-principles model that was developed during the domain engineering study is easily described as M whereas operational data is referred to as O In the construction of digital twins of physical systems, a form of electronic learning known as the 'physics informed' has been deployed. Including physical understanding into knowledge-driven data-based ML models can aid in learning better representations of the problem to generalize to out-of-distribution test data, ultimately reducing training time, and constraining the model with the physical understanding of what it is doing. The reason for this is that in certain circumstances feasible remedies and the numbers that model parameters could assume may are restricted by the unveiled principles. Altogether, there exist six presently recognized basic hybrid framework models, which differ in how they integrate the first-principle model and the data-driven ML method. The first research question is therefore how information technology influences confidence in an organization. This method adds a term to a data-driven model's loss function that rewards the predictions for being compliant with the rules of physics. This has been implemented in several subcomponents of the DT including a PDEs solver component, an inverse learning component, a defect identification classifier, a fragility model including a reduced order model and a hybrid physics ML model constructor. A significant advantage of this strategy based on physics-informed ML compared to other ones is the ability to train independently of the data from physics-based simulations that can be missing.

5.8 Hybrid Modeling Approach for Digital Twin Development [23, 24]

A hybrid modeling approach is the most effective for construction of digital twins, combining both physics-based and data-driven methods. Physics models are great when they come to capturing high depth representations as they are designed to highly mimic the behavior of a system by following the laws that govern the types of interactions that take place between objects. This attention to detail allows for a true comprehension of the physics involved. The disadvantage is that this detail costs—this makes generating such models computationally expensive and time costly for complex systems. Also, some of this knowledge of these physical processes can also be missing. Unlike these, the data-driven models look pretty interesting. Based on using the huge amount of data acquired in the past from sensors and other sources and finding patterns and relationships within the system. Several benefits of approaching from this data-centric perspective. The first advantage of data-driven models is their computational time efficiency and faster development compared to physics-based models. They change easily to new information (as new data is added). But data-driven models do have their own limitations. They often can not answer "why" a system behaves the way it does, missing the explanatory capability of physics-based models. Worse still is their reliability, which is largely dependent on the data that exists, or sadly, does not. The performance of models trained for specific scenarios may suffer in conditions that are much different from the scenario used for training.

The hybrid modeling serves as a bridge to join them together in a best of both worlds approach. It is built upon a physics-based model to begin with, which reproduces the most critical governing equations and the interactions among different subsystem components. The core structure is designed to prevent the model from violating known physical law and to prepare the footing for one to come and grips with the system behavior. But beauty is in the next move. The physics-based model is revised over time with the help of the sensor data and other sources. These are then instead fed into a machine learning algorithm to determine if there are any non-linearities or uncertainties or exogenous factors, that the original equations do not properly cover. This data-driven refinement adds adaptability, a real-world nuance to the model. Calibration and validation are both extremely important after this full model is built. The model is then fine-tuned with the help of real-world data to mimic the actual system performance. This could be modifications of parameters in the physics model, or adjustments to data-driven components. Then the model is thoroughly tested, compared with real data, and checked for conformity. And this way of working helps in continually learning that is the power of this approach. This means that all the data-driven components within the model are updatable, and that the model can be re-trained as new data becomes available to continue to improve its performance accuracy. This loop forms a continuous improvement journey that keeps the digital twin as a lived-out-part of real-life system. This hybrid approach builds upon the wavelength strengths of physics-based and data-driven modeling to provide unprecedented fidelity, generalizability, and interpretability in digital twins. The platform enables breakthroughs across industries, from better managing predictive maintenance in manufacturing, to smart grid management of the electrical network and energy savings in buildings. Hybrid modeling will become an even more important tool in the realization of the DSL and the potential insights that can be derived from it as sensor technology evolves and data analytics capabilities continue to expand.

The Salient Characteristic of Hybrid Modeling

By using hybrid modeling, we combine both of them which results in durable and flexible digital image. So, our jet engine example behaves as following:

Derived from Physics: Fundamentally, the model would be rooted in some sort of physics-based surrogate system that describes the pertinent flow, combustion, and energy conversion fundamentals in the engine. This provides the basis for the fundamental engine behavior and keeps the model in line with the physical laws. Refinement with Real Data: In this step, features are engineered, and a model is developed using the real engine sensor data, which, in sense, is a data-driven model. This big data gathered can be analyzed by machine learning algorithms to discover non-linearities, uncertainties and externals influences that the initial equations might not completely describe. For example, the model may generalize on minute manufacturing extremes or low-end dual core processor performance.

This, however, is where the hybrid model has its magic—it learns, and it grows continuously. Data-driven elements are continuously being updated in real-time as the data sensors in engine operation are collected. The engine model will adapt as the engine behaves over time, allowing for changes in performance that can either improve or decrease.

5.9 Deterioration Modelling

In predictive maintenance employing statistical approaches (V2P) where accuracy is always important when it comes to digital twins and, in particular, degradation modelling. In the case of predictive maintenance, which is also called V2P, degradation modeling is imperative for determining when a physical system might fail. Sometimes, the evolution of the aforementioned physical system is characterized by deterioration processes which comprise physics that cannot be easily explained. In such situations, it is impossible to develop a contextual deterioration model which is based on physical characteristics; it is only possible to develop an empirical statistical model. An empirical based deterioration model can be developed using data generated from the failed units that have been accumulated in the past. In order to explain degradation on a single item as well as the multiple causes of failure, numerous statistical models have been recommended. Deterioration of individual mechanisms has been modeled by many stochastic process models out of which some are: Markov and semi-Markov models, Wiener and Poisson processes, inverse Gaussian and gamma processes, and accelerated life testing (ALT) models. Many models have been proposed for degradation modelling that involve many a degradation mode. These are copula functions, multi-phase stochastic process models, piecewise Markov process models, and system level ALT models [25].

Despite their effectiveness in modeling when physical systems would fail, there are some challenges that need to be met before some of the statistical degradation models can be used on Digital Twins. For example, letter grades herein labeled as A, B, C, and D and other highly abstract discrete states are normally used to represent degradation in a Markov degradation model. Applying the said extremely abstract discrete state models in real-time and online by directly updating them with data obtained from the physical system's sensors, which are often collected continuously, can be challenging. Due to this, creating the P2V link in the context of a digital twin system is a very complex venture. The solution to this problem developed by scientists is the combinations of the described Bayesian model update techniques. Another important issue concerning integration of deterioration models into a maintenance decision-making environment is how to update the models' uncertainty. It was identified that to use prognostic models within DTs, especially regarding high-value physical assets and in safety-critical applications, uncertainty quantification is quintessential.

5.10 Model Calibration and Validation

Once formulated, the models need to be calibrated and validated to ensure their accuracy. Calibration involves adjusting model parameters to match the observed data, while validation checks if the model accurately predicts outcomes for new data sets. Techniques such as cross-validation and sensitivity analysis are often used in this phase [18].

In the realm of digital twin development, model calibration and validation serve as cornerstone processes to ensure the fidelity and trustworthiness of the virtual representations. Model calibration initiates the journey by fine-tuning the parameters and configurations of the digital twin model to enhance its accuracy in mimicking the behavior of the real-world system it emulates. This refinement typically unfolds through a cyclical process, where the model's outputs are meticulously compared against empirical data from the physical system. Through iterative adjustments, the digital twin's parameters are fine-tuned until a satisfactory alignment between simulated and observed behaviors is achieved. This calibration endeavor often involves various techniques such as parameter estimation, data assimilation, optimization routines, sensitivity analyses, and robust uncertainty quantification methodologies. Each of these strategies contributes to a nuanced understanding of the model's intricacies, ensuring that it accurately captures the underlying dynamics of the system it represents.

Subsequently, model validation emerges as the litmus test for the digital twin's predictive prowess. Unlike calibration, which involves refining the model against existing data, validation entails subjecting the model to an independent evaluation against unseen real-world data. This crucial step safeguards against the potential pitfalls of overfitting and ensures the digital twin's ability to generalize across diverse scenarios. During validation, the model's predictions are rigorously compared with empirical observations, employing a suite of evaluation metrics and criteria to quantify its performance. Techniques such as independent data testing, cross-validation methodologies, thorough metrics evaluation, sensitivity analyses, and comprehensive uncertainty assessments are indispensable components of the validation process. By subjecting the digital twin to these stringent validation exercises, organizations can glean invaluable insights into its reliability and robustness, thereby fostering confidence in its predictive capabilities [26, 27].

If a model is still a work in progress or has been developed and created but the performance is not accurate, the models are shown how they should be fixed or modified and validated. On the other hand, Calibration is the process of adjusting model parameters to give expected predicted results, against actual results as in validation you will be able to test how does the model able to forecast for new results. Tools: Cross validation, Sensitivity analysis used in this phase [18].

Two of these, calibration, and validation of the model, are core activities in building the d-twins and are considered fundamental in the process because these provide the basis for establishing the credibility of the model as developed. Next is model calibration, where we tune the configuration and parameters of the digital twin model to allow it to behave more like the actual system or process. This involves, in general, iterating the refinement (usually sequentially), after some number of asynchronous validations of the model results are made with respect to real system data. In other words, the digitized model of the process is being honed down to the sound of the physical system that the simulation is measuring. This calibrative work will consists of data driven parameter identification processes, with model based upation of data with software tools which are complemented with mathematical programming for the optimization and consistent mathematical models, capturing changes in the system and changed responses in the physical system model and quantifiable reliability of the confidence levels. Each individually can contribute a bit more information about the model and the relationships it represents within the system we are examining, improving the validity of the analysis:

Model validation Once this data preparation is complete in terms of creating a digital twin, the model validation turns into a significant test of the accuracy of the model itself. Although calibration tries to improve the accuracy of the model based on the available data, validation allows the model to be evaluated with unseen real world data analysis. This is a classical step that is taken to partially address an issue caused by overfitting and ensure that the digital twin can deal with other cases. The validation consists of testing as well under real world conditions to ascertain that the model performed as well as we had hoped, and that additional metrics and protocols are defined to try to measure that as well. The validation steps include various methods such as cross validation structures, complete metric analysis, testing for independency, sensitivity analysis, and complete uncertainty analysis. Through this growth training and test conditions, the organizations can expose the digital twin to reveal fundamental behaviour knowledge, offering greater confidence in the model's stability and reliability, hence supporting acceptance of the solution's predictions and outcomes [26, 27].

In other words, model calibration and validation are the crucial foundational cornerstones in the construction of DT and the assurance of virtual twins' effectiveness throughout a wide range of contexts. They also actively contribute to predictive maintenance, process improvement, process monitoring and decision making in today's dynamic industrial environment of operations.

5.11 Integration with Digital Twin Framework

At a high level, the integration should describe how the validated mathematical models should be applied in the DT environment with relative ease. This shortly entails defining accesses to the real-time data and ensuring that the models can be run in this environment within the concept of the digital twin. Computational resources management: Large scale computers such as supercomputers and the use of cloud environments. Analogy resulting from such an approach will aid the high-performance computing and cloud-based environment in managing the computational load effectively [28].

(i) *Model Integration:* The mathematical models have been successfully validated and now the digital twins have embedded these within the system in the form that these are available currently—without changing the already established formation. This may involve updating the model such that they can be operated in the context of the digital twin, as well as providing a description of how the other parts of the framework interface with the model.
(ii) *It additionally* brings data into the digital twin framework to help it keep updated (e.g., sensor information, databases, etc.) ...) constantly. They are initialized in the math models which enable the physical system behavior to be emulated and analyzed in real-time [14].
(iii) *In this part,* the proposed integrated mathematical models are used to estimate the responses of the physical system in the context of DT environment. This can include doing projecting, doing assumption checks as well as running operating scenarios and situations.
(iv) *Visualization and interpretation:* The results received in the form of data and performance indicators are presented in simple graphic applications with reports and the display that allows the stakeholders to navigate through the discovery received in the course of deploying the digital twin. In the case of tabular data, a grid system-based dashboard is followed and in case of comparative data 2 D charts are commonly used and in case of spatial data, 3 D models are used.
(v) *Decision Support:* Based on these ideas, the insights created out of the simulations and analysis are incorporated in the framework of digital twin below as supporting decision making. This could be that us organizing they embark on maintenance or process improvements or just inform a stakeholder of threats or immediate opportunities, which the model does.
(vi) *Feedback Loop:* A loop to conform and also to adjust the mathematical models with a more precise type based on the feedback given. Supplemental feedback from actual observation & user interactions to inform changes to such models in the meantime. As such there is a need to ensure that these models are kept current and representative as the system expands and changes over time.

They can introduce authenticated mathematical models via the DT platform, which owns the superior capabilities of digital simulation and analysis to enhance their understanding experience to improve their operation performance providing optimized results which could help in decision making. CFD Simulation Capabilities uses in Digital Twin Framework This section of the article is a perspective view of the overall digital twin structure, digitally representing the data-driven twig statements while including the CFD simulation in Ansys and the systematically managed physics model with the physics method working at a specific range defined by the user.

5.12 Enabling Technologies for Physical to Virtual (P2V) Twinning

Real-world physical parameters and any other information that can be collected can now be introduced into the virtual world. As opposed to the prior scenario where a plethora of real-world data had to be procured, our choice of projective case P2V is stronger. The ability to reason and decide how to secure or avoid threats and risks (V2P2) foresee the outcomes of the physical system by utilizing its virtual representation in real-time. The digital model can be updated by including real-world measurements of its movement or position. Continuing the discussion on various P2V updating mechanisms, we will now shift our focus to "Modelling Methodologies". For instance, utilizing real-time movement data collected from an Internet of Things (IoT) system to construct a digital replica that simulates behavior in the manufacturing process. To identify disruptors, analyze the anticipated location data and compare it with the actual position data obtained from real-world sources. By enabling digital models to move within a virtual shop floor, it becomes possible to optimize various components of both the manufacturing system and shop instruction sheet simultaneously. Mobility data encompasses real-time information on the location and direction of people, cars, and other objects inside the smart city area, along with data on infrastructure and buildings. The concept of digital twin, formerly limited to additive applications, will now be integrated into future city planning and programming for the development of smart cities [19]. By utilizing the empirical data collected within a factory, physics-based systems are developed and evaluated collaboratively anytime an individual makes a contribution. By employing similar strategies across several domains, an original work from each person only signifies that they have created a pioneering masterpiece. By utilizing that method, the differences can be extended or condensed [8].

For example, we can utilize flux data obtained directly from sensors installed on aircraft structures to evaluate the potential risks associated with digital models in the aerospace sector. Surprisingly, when we contemplate abstract concepts, our minds instinctively shift their focus towards the tangible aspects of the actual world. Based on the knowledge provided by industrial consulting, it is possible to achieve optimization within a few decades using physics-based online topologies that can be physically held. Therefore, the emergence of a notion commonly referred to as a digital replica came into being. Due to the availability of multiple hardware-in-the-loop simulation methodologies, an increasing number of industrial organizations are utilizing real-time data to evaluate their systems. It will significantly decrease the cost in terms of both time and money, which is vitally essential for doing these tests.

A large amount of data is required urgently at the utmost speed at this moment. If we consider one perspective, this may be categorized as either wired or wireless. Various wired communication media have varying data rates and distances. The available choices encompass optical fibers, coaxial cables, and Ethernet (twisted pair) cables, among other options. The most efficient method of transmitting data is through a direct physical connection between the source and the receiver. Implementing this will significantly optimize the process, preventing any data loss that may occur during traversal of these paths. NFC (Near Field Communication), Bluetooth, Wi-Fi, and ultra-wide band are just a few examples of the numerous wireless communication protocols available.5G enables instantaneous data acquisition for large and intricate sensor networks in the future. The Industrial Internet of Things (IIoT) is the term used in the industry to refer to networks of sensors and devices that serve as sources for digital twins. In summary, the digital twins discussed thus far utilize the transfer mechanisms to conduct real-time simulations [29].

5.13 Conclusion

This is a deep dive into the modeling of digital twins, emphasizing the importance of mathematical models being capable of reproducing the dynamics observed in the physical world. The chapter opened with an explanation approximately the improvement of mathematical fashions that shape the idea of digital dual technology. Such models are a requirement to properly represent the physical and dynamic world/laws of physical systems in order for the digital twin to make sound conclusions or forecasts.

The chapter also examined some mathematical tools used in diverse arenas. For instance, Computational Fluid Dynamics (CFD) equations play an important role and are required to be solved in simulating the blood flow in the heart, and machine learning algorithms can be used to predict the failure modes in Predictive Maintenance. This provides the right medium to correlate different operating conditions with maintenance requirements with the help of statistical models. By considering some of the mathematical models above, the applicability and versatility of these models in yielding accurate and functional digital twins can be better appreciated.

The last part of the chapter was dedicated to the implementation of validated mathematical models into the digital twin framework. To even start working with React, you need to go through an integration process, which can take many steps to complete.

Integrating Models- Calibrated and validated models need to be compatible with the digital-twin architectural platform. Data Ingestion-Flanging sensor data and historical information in current time through digital twin for the simulation of process with precision. Simulation and Analysis—Running simulations with the connected models to study system behaviour, predict future states, and perform what-if analysis. Visualization and Interpretation—Communicating the results with clear and easy-to-understand dashboards and visual tools. Decision Making means employing the generated insights from a digital twin to make decisions, processes actions and operate optimization.

In summary the modeling of digital twins as discussed in this chapter is no simple mathematical modeling effort; instead, it is a complex process that necessitates advanced mathematical techniques, rigorous calibration and validation and seamless integration into a broader digital twin framework. This comprehensive strategy helps provide resilient, precise and valuable digital twins in different realms including healthcare, manufacturing, urban planning, among others. As digital twin technology continues to evolve, we expect that the principles and best practices articulated in this chapter will continue to serve as a grounding mechanism for creating comprehensive and rigorous digital twins that unlocks the promise of digital twin transformation.

References

1. Tao F, Xiao B, Qi Q, Cheng J, Ji P (2022) Digital twin modeling. J Manuf Syst 64:372–389. https://doi.org/10.1016/j.jmsy.2022.06.015
2. Wright L, Davidson S (2020) How to tell the difference between a model and a digital twin. Adv Model Simul Eng Sci 7:13
3. VanDerHorn E, Mahadevan S (2021) Digital twin: generalization, characterization, and implementation. Decis Support Syst 145:113524
4. Bazaz SM, Lohtander M, Varis J (2019) 5-dimensional definition for a manufacturing digital twin. Procedia Manuf 38:1705–1712
5. Tao F, Qi Q (2019) Make more digital twins. Nature 573(7775):490–491
6. Tekinerdogan B (2023) On the notion of digital twins: a modeling perspective. Systems 11(1):15. https://doi.org/10.3390/systems11010015
7. "Renault Group's vehicles are now built by using a digital twin," Renault Groupe (n.d.) [Online]. Available: https://www.renaultgroup.com/en/news-on-air/news/vehicle-digital-twin-when-physical-and-digital-models-unite/
8. Tao F, Zhang H, Qi Q, Xu J, Sun Z, Hu T et al (2021) Theory of digital twin modeling and its application. CIMS 27(1):1–16
9. Post J, Groen M, Klaseboer G (2017) Physical model based digital twins in manufacturing processes. In: Proceedings of the 10th forming Technology Forum, Enschede, The Netherlands, 12–13 October 2017
10. Møller JK, Goranović G, Brath P et al (2022) A data-driven digital twin for water ultrafiltration. Commun Eng 1:23. https://doi.org/10.1038/s44172-022-00023-6
11. Wunderlich A, Booth K, Santi E (2021) Hybrid analytical and data-driven modeling techniques for Digital Twin applications. In: 2021 IEEE electric ship technologies symposium (ESTS), Arlington, VA, USA, 2021, p 1–7, https://doi.org/10.1109/ESTS49166.2021.9512364
12. Friederich J, Francis DP, Lazarova-Molnar S, Mohamed N (2022) A framework for data-driven digital twins of smart manufacturing systems. Comput Ind 136:103586
13. Tao F, Zhang H, Liu A, Nee AYC (2019) Digital twin in industry: state-of-the-art. IEEE Trans Industr Inform 15(4):2405–2415. https://doi.org/10.1109/TII.2018.2873186
14. A Digital Twin via First-Principles Modelling and Data Analytics for Process Optimisation in Pharmaceutical Manufacturing (n.d.) University of Edinburgh research explorer. [Online]. Available: https://www.research.ed.ac.uk/en/projects/a-digital-twin-via-first-principles-modelling-and-data-analytics-
15. Aristova N, Chadeev V (2023) Mathematical models for digital twins in flexible discrete manufacturing. In: 2023 16th international conference management of large-scale system development (MLSD), Moscow, Russian Federation, 2023, p 1–4, https://doi.org/10.1109/MLSD58227.2023.10303836

16. Krummacker D, Reichardt M, Fischer C, Schotten HD (2023) Digital Twin development: mathematical modeling. In: 2023 IEEE 6th international conference on industrial cyber-physical systems (ICPS), Wuhan, China, 2023, p 1–8, https://doi.org/10.1109/ICPS58381.2023.10128007
17. Ríos J, Staudter G, Weber M et al (2020) Enabling the digital twin: a review of the modelling of measurement uncertainty on data transfer standards and its relationship with data from tests. Int J Prod Lifecycle Manag 12(3):250–268
18. Singh S, Weeber M, Birke KP (2021) Advancing digital twin implementation: a toolbox for modelling and simulation. Procedia CIRP 99:567–572
19. Sun B, Yang C, Wang Y, Gui W, Craig I, Olivier L (2020) A comprehensive hybrid first principles/machine learning modeling framework for complex industrial processes. J Process Control 86:30–43
20. Booyse W, Wilke DN, Heyns S (2020) Deep digital twins for detection, diagnostics, and prognostics. Mech Syst Signal Process 140:106612
21. Chakraborty S, Adhikari S (2021) Machine learning based digital twin for dynamical systems with multiple timescales. Comput Struct 243:106410
22. Chakraborty S, Adhikari S, Ganguli R (2021) The role of surrogate models in the development of digital twins of dynamic systems. Appl Math Model 90:662–681
23. Zheng Z, Ni D (2022) A hybrid modelling approach for the digital twin of device fabrication. In: 2022 China semiconductor technology international conference (CSTIC), Shanghai, China, 2022, p 1–3
24. Wagg D, Worden K, Barthorpe R, Gardner P (2020) Digital twins: state-of-the-art future directions for modelling and simulation in engineering dynamics applications. ASCE-ASME J Risk Uncertain Eng Syst Part B Mech Eng 6:030901
25. Davila Delgado JM, Oyedele L (2021) Digital twins for the built environment: learning from conceptual and process models in manufacturing. Adv Eng Inform 49:101332
26. Schroeder GN, Steinmetz C, Rodrigues RN, Henriques RVB, Rettberg A, Pereira CE (2021) A methodology for digital twin modeling and deployment for industry 4.0. Proc IEEE 109:556–567
27. Bhatti G, Mohan H, Raja Singh R (2021) Towards the future of smart electric vehicles: digital twin technology. Renew Sust Energ Rev 141:110801
28. Platenius-Mohr M, Malakuti S, Grüner S, Schmitt J, Goldschmidt T (2020) File- and API-based interoperability of digital twins by model transformation: an IIoT case study using asset administration shell. Futur Gener Comput Syst 113:94–105
29. Minerva R, Lee GM, Crespi N (2020) Digital twin in the IoT context: a survey on technical features, scenarios, and architectural models. Proc IEEE 108:1785–1824

Chapter 6
CyberTwinAI: Enhancing Security and Efficiency in Industry 4.0

6.1 Introduction

Industry 4.0 has brought unparalleled change in manufacturing and the Industrial Revolution which is drastically changing the outlook of things. The fourth industrial revolution entails digital automation across all industries within the economy, and there has been creativity as well as productivity in existence. However, with the conveniences that adapt to the new era's framework, come the inherent challenges of security and efficiency of operations. Based on this premise, this chapter begins a journey to discover how synergism derived from advertising technologies such as CyberTwinAI helps to address these important problems and drive Industry 4.0 and put it into a time frame within a plan or a year or a month or within a breath, it results into a period of capability and resilience like which has never been heard of before. Industry 4.0 mark represents a major step forward in thinking among different companies to evolve the way production and industries are created. The basis for this is just automation, data management systems, and intertwined platforms that mean robots talk, adapt, and learn. However, in this setting, effectiveness, and safety are crucial aspects of consideration. More stringent security measures are expected, needed, and demanded because the physical and digital environments, assets, and identities are becoming increasingly blurred and are open to cyber-attacks [1]. In this e-revolution, efficiency is at once the motivating principle for competitive positioning, eliminating waste, idle time, and operating costs.

In light of these demands, allow us to introduce CyberTwinAI. At its core, CyberTwinAI is the fusion of two crucial technologies: Cyber Twins, which have a digital replica or a twin of the physical structure of an asset, and Artificial Intelligence also known as AI a force that can do anything when it comes to data analysis and decision making. On which the CyberTwinAI manages security and concern for productivity, Industry 4 revolution relies on these aspects. It is crucial to discuss that the combination of 0 presents is the closest to the ideal convergence of these technologies. This chapter explores this idea to the very essence, offering

© The Author(s), under exclusive license to Springer Nature Switzerland AG 2025

S. Gupta et al., *Digital Twins*, https://doi.org/10.1007/978-3-031-76564-3_6

comprehensive coverage of this subject matter, its capacity to reinvent the structures of companies, and its operations in the industry 4.0 ecosystems [5].

6.2 The Challenges of Industry 4.0

Industry 4.0, commonly referred to as the Fourth Industrial Revolution, represents a shift in the usage of technological tools and processes in companies. Thus, it can be defined as the utilization of digital technology, data, and smart systems within production and other industrial applications. Industry 4. 0, often referred to as the fourth industrial revolution, originated in Germany and is also tightly connected with the idea of a smart factory or 'Smart Industry. 'In its essence, Industry 4. 0 is to create a new high level of industry automation, intelligence, and networking to have such smart systems and equipment that those devices and tools coordinate and communicate with each other has been never seen before [2].

6.2.1 Key Ideas for Industry 4.0

There are a few key concepts on which Industry 4.0 is based, which give credence to the potential transformational power of this new era in manufacturing technology. This includes:

- Interconnectivity: Everything is connected to everything else in an Industry 4.0 manufacturing environment. IoT, for instance, and other such technologies are put to use so that machines, sensors, and such systems can communicate in real-time.
- Information Transparency: Digitalization creates transparency in systems and processes. Data is collected, communicated, and analyzed compulsorily to reflect operational insights in real time.
- Technical Support: On the one hand, today's intelligent systems and technologies of machine learning, and artificial intelligence, support and empower humans in making their decisions and taking their actions.
- Decentralised Decision-Making: There will be less need to avail performances of central control since machines and systems can decide on their local level of view.
- Virtualization: Digital twins among other virtual copies of a process are used in mocking and perfecting real-life procedures.

6.2.2 Challenges of Industry 4.0

With all the benefits, Industry 4.0 has some points to note in terms of issues [7].

- Cybersecurity: System vulnerability and data prevention are very much probable with increased interconnection.
- Data Management: This new industry symbiosis generates gigantic data; this needs to be managed efficiently, stored, and analyzed.
- Skilled staff: The management requires workers who can efficiently manage and maintain the latest technology such as robotics and artificial intelligence.
- Cost of Implementation: The implementation of technologies associated with Industry 4.0 is relatively expensive in the short term. The initial financial outlay may pose certain challenges for businesses.
- Ethical Issues: The more machines make decisions, the larger those issues become-including questions of responsibility and guilt.

This understanding of such barriers is going to be important for the establishments that are navigating through such complexities of Industry 4.0 and looking for innovative solutions like CyberTwinAI to sort out the security and productivity challenges.

6.3 Industry 4.0's Growing Need for Cybersecurity and Efficiency

Efficiency and cybersecurity are becoming more and more critical in this competitive environment of Industry 4.0. Since digital technologies, automation, and data-driven operations have taken over, organizations are exposed to a more significant possibility of cyber-attacks, and the need for more excellent security can be felt. Interconnection of systems, growth of sensors and devices, and continuous flows of data make it easier to stage a potential cyberattack, data breaches, and unauthorized access. Because of the extremely high volume of sensitive data and valuable operational data at stake, there certainly never was a more critical time to secure these modalities of digital environments. Industry 4.0 organizations-manufacturing, logistics, or otherwise-must afford proper cybersecurity sufficient prioritization in securing the assets, maintaining customer trust, and assurance of business continuity [1].

At the same time, Industry 4.0 has operational efficiency as its major goal. Corporations analyze their operations due to the stiff competition in the global market, as well as to optimize resources, minimize waste, and thereby boost production. Productivity, resource efficiency, and decision-making ability are prominent elements of success in the current digital age. Capability of data-based decision-making, optimizing operations, and reduction of downtown. Like in any other industry, Industry 4.0 has shifted from data-driven and thus reactive approaches to ones that focus on the bottom line in proactive ways. Therefore, the race to maximize efficiency has become a catalyst pushing companies to fully exploit digital technology and innovative solutions like CyberTwinAI if they are to remain resilient and competitive in ever-changing industrial landscapes [2].

6.3.1 CyberTwinAI: Theory and Structure

The concept of CyberTwinAI emerges as an innovative creation of the technological revolution that is taking place in the middle of Industry 4.0 fully prepared to transform and make perfect the way human beings interact with both the physical and the digital world. At its very core, CyberTwinAI is the synchronization of two vital technologies: Artificial Intelligence, or AI, and Cyber Twins, all elements that form a dynamic and revolutionary platform for organizations and industries to grow in this new century.

The concept of the CyberTwinAI is based on the concept of Cyber Twins. It means digital duplicates or look-alike of actual capital assets that could be anything from equipment to finished goods, to an entire manufacturing plant. Digital twins represent not just the physical attributes, but the history of performance and operations of their physical twin counterparts dynamically in real-time. Due to the incomparable levels of understanding, tracking, and predictive ability available through the use of this digital twin, businesses would be able to achieve knowledge as well as make defensible decisions that could be fact-based [8].

Intelligent artificial systems (AI): Cyber Twin is powered by Artificial Intelligence horsepower. Cyber Twins have embedded AI in various forms from Machine learning to deep learning and beyond to analyze data, discern patterns of information, and render intelligent decisions. Powered by artificial intelligence, AI, Cyber Twins are more than dumb digital duplicates. They are now systems of cognitive learning that understand the world around them, predict events, and make decisions independently.

Consequently, the CyberTwinAI framework brings together these two fundamental components to have a symbiotic interaction of the digital twin with its AI-driven cognitive capabilities. From the practitioner perspective, this means AI-enabled Cyber Twin actively participates in industrial events to optimize monitoring; provide maintenance prediction, and resource employment optimization, and, contribute to the main goals of Industry 4.0 with more security, efficiency, and competitiveness. This will be elaborated in more detail in the following chapters, which explain how CyberTwinAI uses this framework to transform industrial operations in the context of Industry 4.0.

Figure 6.1 shows the conceptual representation of CyberTwinAI; the left side is the world of practical physical assets, sensors, operation data, and past performance; and on the right side is the digital world, comprising Cyber Twins, historical performance data, real-time data from sensors, and AI algorithms for data analytics and decision-making. CyberTwinAI implements the concept and framework for the improvement of industrial operation activities and decisions; the arrows indicate the data communication and interaction between the physical space and the digital twin.

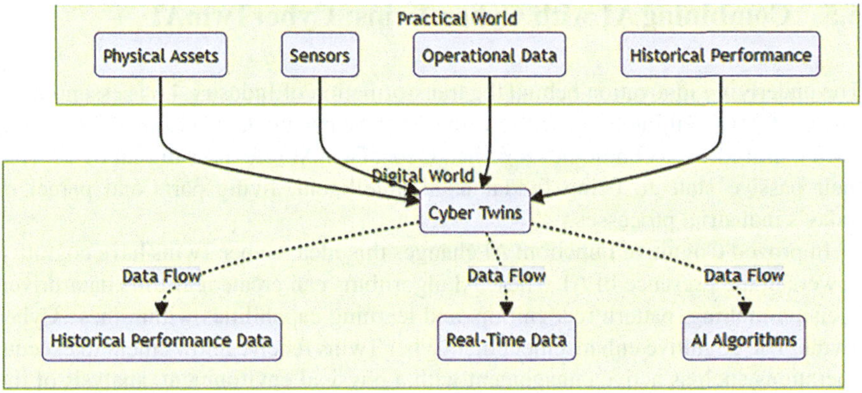

Fig. 6.1 Structure of Cyber Twin to improve the industrial process

6.4 Artificial Intelligence (AI) and Cyber Twins: An Analysis

Artificial intelligence is the vanguard of technology in this era of Industry 4.0, and indeed, it is changing all kinds of industries. Artificial intelligence, in simple words, is the creation of computer systems that can think, learn, and act just like human beings-that is, systems with embedded computer intelligence like understanding natural language, recognizing patterns, learning from experience and data, or making decisions. The concept of cyber twins, when merged with AI technologies like deep learning and machine learning, can bring about complete change in the industrial sector.

AI algorithms embedded in the Cyber Twins could analyze massive amounts of historical performance data and real-time data, and complex patterns, anomalies, and trends that may be hard for human beings to discover can be identified by them. The ability to derive meaning from data is what industries must acquire to improve operational efficiency, forecast maintenance, and be more data-driven in decisions. "Activating" Cyber Twins with AI empowers them to take a much more active role in decision-making: they can self-optimize operations; they can change their parameters on their own; and, at best, they can even begin avoiding problems-be they sidestepping an upcoming malfunction or simply dodging excessive energy consumption and/or wear and tear on equipment. The assumption here is that CyberTwinAI Wiley combines AI and Cyber Twins or digital twins, in effect creating a dynamic framework for changed and revolutionary improvement of security and efficiency but also opens up whole new possibilities in the industrial sector.

6.5 Combining AI with Cyber Twins: CyberTwinAI

The underlying innovation behind the transformation of Industry 4.0 is essentially a fusing of Artificial Intelligence, AI with Cyber Twins, or "CyberTwinAI." In bringing a multitude of technologies together, CyberTwinAI refocuses digital twins from their passive state to being fundamental, intelligent, living parts and parcel of today's industrial processes.

Improved Cognitive Function: AI changes this idea. Cyber Twins have cognitive powers in the presence of AI. These AI algorithms can create real-time data-driven decision-making, pattern recognition, and learning capabilities within these Cyber Twins. The cognitive enhancement that Cyber Twins receive allows them to execute operations such as active engagement with a physical environment, analysis of the data coming in, and dynamic response to situations as they are changing—among others—on top of just observing and replicating.

1. Predictive Analytics: The uses of such models make AI integrating with Cyber Twins very useful when giving you predictive analytics. It helps to predict the further materialization of patterns, deviations, or issues in the future, using historical and current data. For example, a manufacturing machine's cyber twin can utilize the data gathered from the sensors to determine when a mechanical breakdown would happen, and thus facilitate the taking of precautionary measures. With regards to its application in cybersecurity, a network's Cyber TwinAI can detect and assess data traffic irregularities because, on its own, the Cyber TwinAI has inherent abilities in predictive analytics and reconnaissance of breaches. Implementing the measures advocated in this paper will cut down on downtime, optimize resource utilization, and enhance overall effectiveness, hence transforming several sectors thanks to predictive analytics.
2. Autonomous Decision-Making: In Cyber Twins AI minimizes the functioning of one from the other and allows each of them to act on their own. The decision-making capability of Cyber TwinAIs enables them to start tasks autonomously when they observe deviation from norms or an impending issue. For instance, a Cyber TwinAI of an autonomous car may decide to alter the path it selects to increase travel efficiency, using info on traffic and weather patterns. A Cyber TwinAI can adjust the efficiency on a production line because-parameter of a given machine can be tweaked. It also means that the requisite rapid responses to fluctuating conditions can now be achieved because of these self-sourcing decision-making capabilities, which later enhance efficiency and resources.
3. Cyber TwinAIs can learn and evolve and can be recharged with fresh knowledge anytime. They can improve their decision-making with the help of continuous training in machine learning and continuous data availability. Here, the student is built up to be able to detect patterns explanatory within diverse territories as they become more accomplished in arriving at decisions with particulars. This is particularly useful when an operational environment that the business relies on transforms over time.

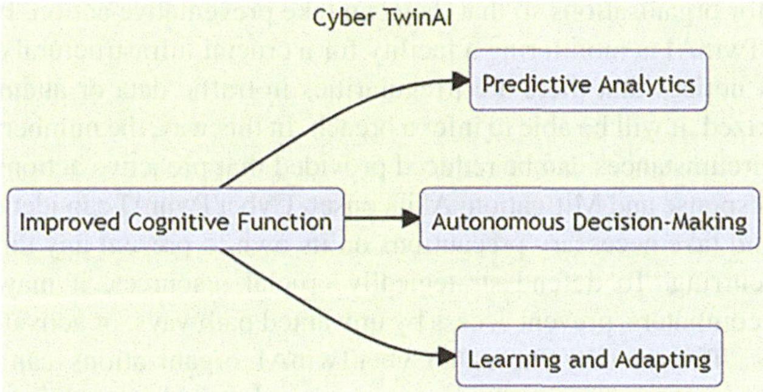

Fig. 6.2 Cyber TwinAI

Figure 6.2 shows the combination of AI with Cyber Twins to form CyberTwinAI. It not only improves efficiency and security but also opens the door to intelligent, self-governing, and flexible systems, which are essential for success in the Industry 4.0 environment. An era of industrial processes that are more robust, responsive, and optimized is ushered in by Cyber TwinAI's ability to do cognitive, predictive, and autonomous tasks.

6.6 Strengthening Security with CyberTwinAI

In the context of 4.0, the flexibility achieved in production and related processes can be explained by the following reasons. As we now see digital ecosystems and networked systems dominating the industrial environment, it is cybersecurity, with a value of 0. Threats and weaknesses that can be targeted by cyber threats and attacks have multiplied due to the Internet of Things (IoT), connected technological systems, and data. Taking into account these challenges, security is central to CyberTwinAI in pursuit of the notion in this setting.

1. Real-time Threat Monitoring: No matter it is an industry that produces goods, a logistics network, or a utility infrastructure, CybetTwinAI creates or acquires a depiction of a physical subject which it calls "Cyber Twin," which is an accurate representation of the physical counterpart in real-time. Real-time as such may help in the continual monitoring of the physical system and its digital replica. CyberTwinAI is particularly effective in finding any unusual activity that can potentially make an organization vulnerable to security threats. Subsequently, CyberTwinAI has the potential to flag various issues as and when they occur; for example; unusual data flow, attempts at data breach, or system discrepancies.
2. Predictive Analysis: In logic, CyberTwinAI employs AI predictive analytics alongside far more conventional measures of monitoring to proactively identify security threats before they can arise. It means that it can predict potential risks because with the help of evaluating past data one can see trends which is kind of

an alert for organisations so that they can take preventative action. For instance, if CyberTwinAI is monitoring a facility for a crucial infrastructural corporation, then if it notices that there are irregularities in traffic data or attempts that are unauthorized, it will be able to infer a breach. In this way, the number of security-related circumstances can be reduced provided that proactive actions are taken.

3. Quick Response and Mitigation: At its onset, CyberTwinAI can identify security threats and take necessary precautions on its own to prevent any threat or harm from occurring. To defend strategically crucial resources, it may quarantine affected computers, prevent access by unwanted pathways, or activate protection measures. This is why through CyberTwinAI organizations can respond to threats in a matter of seconds not to mention that with the AI systems performance, it will act far more decisively and quickly than human response time measured in minutes and hours is far too slow.
4. Ongoing Education and Adjustment: Another advantage of CyberTwinAI is that it is capable of learning in the process and with time the AI system becomes healthier. It improves the capability of discovering new threats by machine learning and deep learning approaches. It develops to enhance the discovery of new mannerisms, attacks, and vulnerabilities to strengthen general security with time.
5. Cost-effectiveness and Scalability: Thus, CyberTwinAI offers organizations an effective security solution that can be easily implemented across different levels. This is because it can be applied across an organization from its digital core to its edges and is hence appropriate for complex systems. Also, the ability to monitor and report results in decreased reliance on traditional methods that require significant human input and hence it is cost-effective.

6.7 Industry 4.0 Security Challenges and how CyberTwinAI Can Help Secure and Monitor Industrial Systems

Figure 6.3. Specifically, Industry 4.0 has specific security challenges that are outlined below

- The Breach of Data: With the increasing integration of such technologies into people's everyday lives, cybercriminals have a field day given the massive amounts of data generated by these technologies. Confidential information such as trade secrets, patents, and other intellectual property, and other corporate proved information may be at risk of being leaked by hackers.
- Cyberattacks: Systems Operating in Industry 4.0 are prone to several cyber threats including viruses, malware, ransomware, and DDoS attacks. Such attacks can lead to data degradation, interruption of operations, and loss which are inconducive to the survival of these companies.
- IoT Vulnerabilities: Industry 4.0, "The aims of this study are to; Identify and classify the information technologies used in Industry 4. 0 largely works with Internet of Things (IoT) devices; however, the latter are often poorly protected

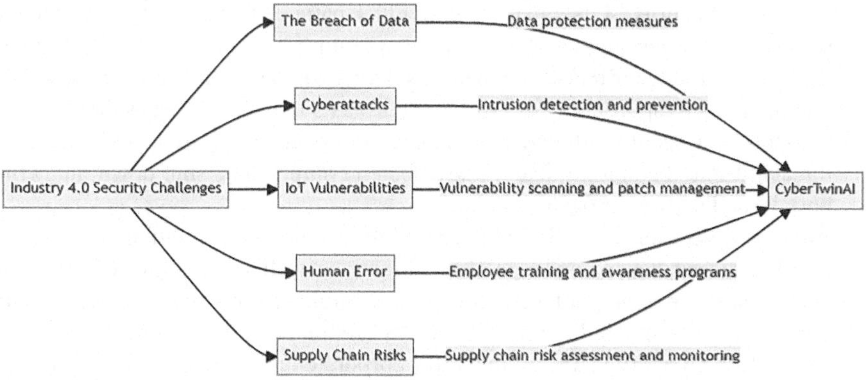

Fig. 6.3 Industry 4.0 security challenges

from cyber threats. These devices have loopholes through which an attacker can launch and perform exploits to gain unauthorized access to the networks.
- Human Error: Based on the current advances in technology and the rapid expansion of individuals' use of the devices, chances of making mistakes and producing undesirable outcomes are also higher. Employees can be reckless with data open unknown emails with dangerous consequences or accidentally set up systems wrongly.
- Supply Chain Risks: To establish organizations, that suffer the risks and security vulnerabilities resulting from linkages' weakness in the supply chain, organizations have been classified.

6.7.1 How CyberTwinAI Can Monitor and Secure Industrial Systems

Security issues and problems in Industry 4.0 are well-managed through CyberTwinAI security.

- Real-time Monitoring: When CyberTwinAI is coupled with the digital twin as in the discussion, it becomes possible to always oversee both the digital and the physical aspects of the industrial systems. To identify anomalies from the expected, it constantly audits the data passing through it, the characteristics of the system, and the output of the sensors.
- Anomaly Detection: CyberTwinAI can review massive amounts of data to find irregular patterns that signify a security threat via the application of AI. The benefit of this proactive approach is that declining cessations can be identified at an early stage.

- Predictive Security: First, like an individual patient, CyberTwinAI prognoses security threats before they occur. It predicts future risks or next attacks through analysis of past occurrences and patterns, which enables one to prevent attacks.
- Autonomous Response: CyberTwinAI controls its talk which means that when it recognizes a security threat, it can respond proactively. To quickly contain threats, it may be able to quarantine affected components, shut down unlawful portals, and activate protective elements.
- Adaptive Learning: CyberTwinAI is a perfect example of something that is constantly gaining new knowledge from threats as well as adapting to different approaches to reaching a target. Adaptive learning ensures that the system will remain effective in handling new threats by embracing what is new.
- Scalability: CyberTwinAI is suitable for corporate organizations and businesses with diverse and complex digital structures because it can be integrated into different industrial platforms. That feature makes it versatile in its ability to ensure that the package offers comprehensive security.

6.7.2 CyberTwinAI Applications in the Real World to Improve Security

Following are some real-world uses of CyberTwinAI in the context of Industry 4.0 to improve security [3]:

- Secure Smart Manufacturing:

- CyberTwinAI is deployed in a smart manufacturing facility with the structure of the production line and machines. This digital replicate employs data from integrated sensors to always survey the natural state of physical space. CyberTwinAI can execute plans autonomously to counter a threat if there is suspicion of fraud, for instance, or intrusion in the network. For instance, it may lock the corresponding computer or alert security personnel about the need to attend immediately and prevent such cyber-attacks or intrusions into the production line.
- Grid Security for Energy:
- Energy security can be regarded as an essential problem, especially in smart grids where vital infrastructures must be protected. Both ETBO and CyberTwinAI are utilized to mimic elements of the energy grid. It monitors; the production, distribution, and consumption of power it does this continuously. The technology can independently initiate countermeasures, like blocking specific segments of the grid, redirecting electricity, and alerting security personnel, if the technology identifies something that looks like a cybersecurity threat or intruder activity that might be indicative of a cybersecurity breach of power infrastructure.
- Supply Chain and Logistics Security: Several shipments and warehouses are involved in the supply chain and logistics sector and it is vital to protect the network. This framework allows the creation of a CyberTwinAI that mimics sup-

ply chain procedures. The calendar is used to keep track of the delivery dates, stock status, and transport routes. The CyberTwinAI can autonomously trigger an alarm and take action for instance in cases where the AI algorithm detects a violation of rules like tampering with shipments or diversion from planned routes, the CyberTwinAI can cause a diversion of such shipments or close down areas that are prohibited to access.

- Smart Building Security: Enabling security in smart buildings, CyberTwinAI twins every system in a building, including environment control systems, surveillance cameras, and access systems. If it detects any odd pattern of behavior, say, unauthorized access or an attempt to tamper with surveillance equipment, CyberTwinAI may independently trigger several responses, including shutting down access points and raising alarms, while informing security with the same speed.
- Healthcare Security: The priority of every healthcare industry is to keep its patients' information and medical equipment safe. CyberTwinAIs can simulate the healthcare network and monitor network traffic, medical device usage history, and accesses to patient records. It enables the CyberTwinAI to rapidly secure patient data by closing unsolicited access points and alerting IT and security staff to detected events of suspected data breaches or cyber-attack incidents [6].
- Vulnerability Assessment and Management: With CyberTwinAI, one has the ability to digitally reproduce any IT infrastructure. The digital twin can be used to continuously assess and discover vulnerabilities within the system. The security teams could work on fixing or mitigating the issues before an attacker in the physical world exploits them.
- Anomaly Detection: CyberTwinAI can help in anomaly detection by continuously monitoring the behavior of digital twins against predicted behavior to detect a possible security compromise, enabling its early identification and requisite action.
- The incident response simulation feature of CyberTwinAI is intended for simulating cyber-attacks and their consequent behavior across the digital twin of a network within an enterprise. This shall enable security teams to exercise their incident response plans in readiness within the real world against these threats.
- Education and Training: One way in which CyberTwinAI can apply security experts is by improving the general training. Trainees can learn to type out responses to threats in a trainer-like environment, having controlled environments used as simulations.
- Security Patch Testing: This will mean, with CyberTwinAI, one can apply it in predicting how security updates and patches designed to be performed on actual systems would influence the digital twin. This makes it less probable that patches wouldn't have some unseen issues, complications, or other vulnerabilities.
- Phishing and Social Engineering Simulations: This would test the workforce's knowledge in an organization and how they might react when faced with such threats. CyberTwinAI runs this simulation to harden the valued security measures usually people-centric.

- IoT and OT Security: any gadget of the so-called Internet of Things and industry operational technologies can be created with CyberTwinAI as realistic virtual replicas for close monitoring and protection of critical assets essential to support national missions.
- CyberTwinAI can be leveraged to run supply chain security analysis for systems and leave surface vulnerabilities or threats that might impact the security of their systems within the company's supply chain.
- Regulatory Compliance: In the context of compliance, there is a possibility that CyberTwinAI may help organisations ensure they maintain legal and regulatory requirements set by industry-specific laws and policies.
- Threat Intelligence Integration: Instead of virtual modeling as a simulation, threat intelligence data can also be fed through to improve their performance in addressing emerging threats; though reduced to real-time.
- User Behaviour Analytics in CyberTwinAI focuses on how the users perform and interact with the digital twin model to identify any abnormal or unusual behavior that may suggest insider risks.
- Real-time Network Defence: The intervention is possible due to CyberTwinAI, which can also limit the threat surface in real-time by blocking or isolating the malicious network traffic or activity. This is made possible by AI and machine learning.

CyberTwinAI, a concept related to creating digital twins of real-world systems for cybersecurity purposes, has several applications as indicated in Fig. 6.4 in the real world to improve security. Here are some ways in which CyberTwinAI can be applied to enhance security.

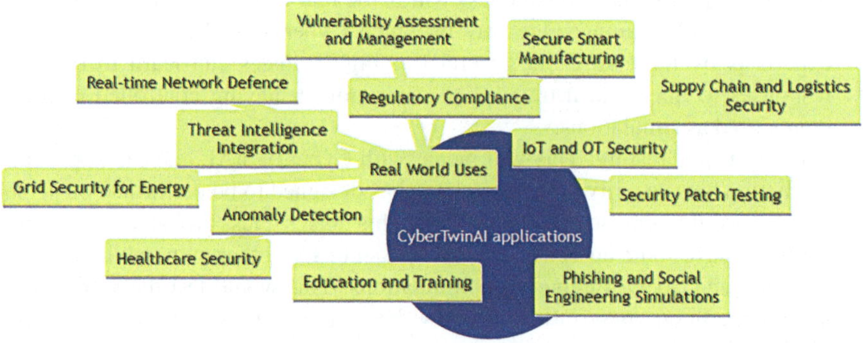

Fig. 6.4 Applications of CyberTwinAI

6.7.3 Leveraging CyberTwinAI to Increase Efficiency

Energizing operational efficiency is a pressing sector goal for sectors in the Industry 4.0 space. In this regard, CyberTwinAI emerges as the disruptor in this premise, challenging established ways through which corporations and businesses enhance and optimize their processes to boost yields while eliminating wastage. Thus, CyberTwinAI offers a blend of AI and digital twins as a tool to address the issue of efficiency enhancement from a diverse perspective.

The primary functions of CyberTwinAI are real-time monitoring and predicting events and outcomes. If constantly observing and collecting knowledge from real systems which it mimics, it looks for patterns in data streams. Using these predictions, it can then predict future problems that are such as equipment faults, manufacturing choke points, or mechanical breakdowns. Due to this, preventive measures and proper acquisitions can be well-planned thus reducing downtime and ensuring optimization of available resources. However, what makes CyberTwinAI stand out is that it is equally able to alter operational settings on its own about given circumstances as a way of trying to efficiently integrate methods. To avoid loss of resources and to gain optimal output, a smart factory, for instance, can alter the production program the parameters of a specific machine, or even the amount of energy used within a manufacturing plant depending on the overall demand of a given time. Finally, especially the learning and adaptive systems and the option of making data-driven judgments place CyberTwinAI in the center of the efficient processes of Industry 4.0. [4].

6.7.4 Efficiency in Manufacturing and Industry: Its Significance

For manufacturing, and particularly for industries, the key ingredient is being operationally effective. Industry 4.0 This is so especially given the highly competitive global environment that is experienced by organizations today, where more and more is expected to be achieved using fewer resources. Here are some reasons why efficiency is so crucial: Here are some reasons why efficiency is so crucial:

- Cost Reduction: The main consequence of improving the efficiency, that can be noted is that the costs will decrease. Eliminating and avoiding wastage, optimizing processes, and utilization of resources are proactive measures towards decreasing operation costs, some of which are crucial to coverage of costs to retain financial viability and sustainability.
- Enhanced Productivity: It is very important to define efficiency since productivity rises with efficiency. It assists organizations to generate more by the usage of the same or lesser quantities of resources, which boosts throughput and revenues.

- Less Downtime: In industry and especially in manufacturing, any company's unplanned maintenance time is a real drawback. Reliable mapping of real-time monitoring and introducing predictive maintenance together with venturing to optimize operations can slash downtime by so many factors and ensure that there is a constant flow of production.
- Resource Optimisation: Overall, it means that resources are best managed when being used efficiently. This is the machinery, the workers—their skills and strength—the power supplies, and the materials that go into the production of the goods. The effective management of these invaluable assets enables the enhancement of achieving the best possible utilization of the resources whereby students can learn through positive, vigorous, and indeed pleasant experiences.
- Enhanced Quality: It is quite common that quality and an increase in efficiency are two variables that correlate with each other. An efficient procedure in the manufacturing of a product leads to the production of better-quality products, lesser generation of waste, and enhanced customer satisfaction due to fewer defects and variations.

6.8 How CyberTwinAI Boosts Productivity, Cuts Downtime, and Optimises Processes

CyberTwinAI is a revolutionary technology that can enhance industrial and manufacturing productivity:

- Predictive Maintenance: CyberTwinAI always implements active sensing of real-time data from the surroundings and then constructs a digital replica of physical assets. It remains useful in predicting the maintenance requirements since it recognizes anomalies, which suggest that a breakdown is nearby. In effect, since the CT scan machines are used to perform important tasks within the health Germany organizations, Germany Health Care gets to benefit from reduced 'planned' downtimes and increased 'unplanned' uptimes.
- Another unique aspect of CyberTwinAI is that it has an interacting and self-optimizing feature that enables the tool to adjust working parameters on its own after processing data. For instance, an SHM known as a smart factory may adjust the settings of a machine in real-time, adjust the timetable for production, and alter the allocation of resources in response to market demands. This ensures that things remain set and aligned to new environments to meet the set objectives.
- Resource Management: CyberTwinAI has also adopted the best practices when it comes to matters concerning resource utilization. It helps the firms in achieving optimal utilization of labor, energy conservation, and minimization of wastes which help in reduced expenses as well as enhanced production.
- The use of CyberTwinAI also makes it easier for fast and informed decision-making since the machine makes faster decisions than human beings. Zander suggests that, as a result of this company's fast decision-making, 'its operation is incredibly efficient with very few processes disrupted'.

- Ongoing Learning: The model used in CyberTwinAI, the query learning capability, acquires knowledge and evolves from the data it analyzes, constantly. Namely, this means it becomes more efficient with time, as it sees opportunities for optimization or adapting the practice according to new conditions.

Essentially, CyberTwinAI uses digital twins and AI to improve resource management, decrease downtime, and optimize operations to increase productivity. It provides a flexible and dynamic framework that is essential for success in Industry 4.0, where effective operations are required rather than just a competitive advantage.

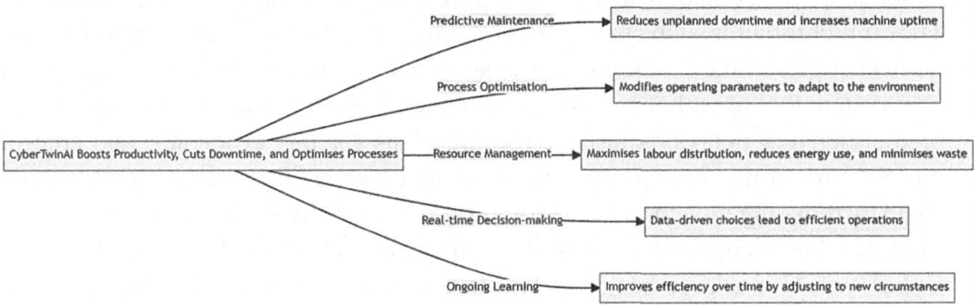

6.9 Case Studies Where Organizations Have Used CyberTwinAI to Their Advantage

CyberTwinAI is another concept that has not been fully developed yet; however, the examples of its applications and cases also point to increases in productivity for various industries. Bearing in mind the purpose of using CyberTwinAI in this book is to demonstrate how different industries can enhance productivity, these cases reveal the following insights about the effectiveness of utilizing CyberTwinAI. It is worthy of note that as a concept Industry 4.0 Digital twins with AI analytical and real-time decisions have helped companies enhance their worth, resulting in reduced downtime and better utilization of resources. These changes have all pointed to the achievement of the overall organizational goals.

1. CyberTwinAI was applied in production facilities as reported by a top automotive firm when doing a manufacturing and predictive maintenance study. The Outline of predictive maintenance skills was predicated on the creation of frequency-specific Virtual models of manufacturing plants and equipment and integrating them with artificial intelligence factors. Safety evaluations, quality analysis, real-time machine health monitoring, failure prediction, and maintenance schedules were performed by the system before any system failure. This they achieved through a marked cut on the unexpected downtimes which they noted were by 30%.

2. Grid and energy management: CyberTwinAI the company applied the solution in a large utility firm with a focus on grid management. The lively data processing was made possible by Preprocessing functions and AI algorithms and by developing virtual models of power grid structures. To maintain the stability of the grid, the CyberTwinAI might have to self-adjust the power flow independently, so that any disturbances such as overloading or failure in equipment may be avoided. Customers would always gain power due to the reduced losses and enhanced total system distribution.
3. Logistics and Supply Chain Optimisation: A logistics organization of the multinational space worked with CyberTwinAI to improve supply chain operations. They might monitor such changes in real time and agree to adapt by creating digital twin copies of their warehouse, transport paths, and inventory management. A new system that uses the technology of artificial intelligence (AI) was adopted to predict delivery delays, adjust the routing plans for fuel efficiency, and control internal inventories in real-time. They pointed out that this enhanced the supply chain by reducing fuel costs by 20% and shortening delivery times by a good 15%.
4. Real-life knowledge on patient flow management in health care: Applying CyberTwinAI in a large health system Patient flow assignment and resource utilization were optimized in several hospitals of a big health care network using CyberTwinAI. In real-time, it was necessary to track patient admissions, room occupancy, and staffing levels, and for this, digital twinned copies of each facility were created with AI algorithms integrated into the system. This allowed them to minimize patient waiting time, get the highest utilization of all the rooms and staff, and effectively address the needs of patients by using Data Mining to predict which resource will be needed and when. This helped in improving patient satisfaction and healthcare quality efficiency in the health sector.

6.10 Challenges and Their Concerns

There are Challenges and concerns associated with CyberTwinAI in Industry 4.0. Despite the technology's enormous potential, several important problems still need to be resolved [8]:

1. Privacy and Security:

   Challenge: Furthermore, CyberTwinAI relies on data acquisition and analysis of Industrial systems and processes. The following issues are crucial the fact that the data is often personal and sensitive Privacy and security issues are always at stake.
   Concern: Lack of data security policy measures might lead to data disclosure and unlawful access and possibly compromise private or confidential and sensitive data. The need for protecting data during creation, collection, usage, storage, and disposal also demands effective cybersecurity measures to be adopted.

2. Ethical Points to Take into Achieve:

   Challenge: When CyberTwinAI is integrated into decisions, issues of morality creep in since it will be involving itself with critical decisions. For example, self-decisions might have consequences for the allocation of resources or the safety of people, which raises the problem of justice and accountability.
   Concern: The key challenge for technology's application is to ensure it works within the parameters of ethical norms and legally determined legislation, thus the need for ethical standards and regulation. When AI is used in areas such as care, privacy, and confidentiality of patients and information at times patients may not willingly give their information willingly also comes into a picture [6].

3. Integration Difficulty:

   Challenge: The implementation of CyberTwinAI in current existing industrial systems may pose some problems such as costs and challenges in integration. It may often be required to integrate into new systems or contort the older ones and this is time and energy-consuming.
   Concern: They must ensure they have allocated enough cash for the merger from the organization's budget and properly plan how to implement the merger. Failure in the management of integration complexity may also lead to possible interruptions in continued operations.

4. Lack of Standardisation:

   Challenge: While CyberTwinAI is still a relatively new field, there are no existing numbing practices or industry benchmarks. Due to this, business entities may encounter challenges in determining the optimal technological combination to adopt as well as the compatibility that comes with it.
   Concern: Since there may be no well-defined and set standard procedural requirements in the enterprises, there might be some issues especially when it comes to harmonization and data exchange between numerous CyberTwinAI systems. The goal of the industry should strive to be better and conform to the international best practices for accreditation which would also help in formulating best practices for the industry.

5. Skill and Training Gaps

   Challenge: The next very important factor that organizations may find pretty hard to overcome is the fact that getting the exact know-how and skill for implementation and usage of CyberTwinAI systems might not be very easy in some cases.
   Concern: To overcome this issue the company, will need the services of other professional people so that it may either buy other training programs to enable its employees to gain more knowledge. The chances of successfully putting CyberTwinAI into action could be offset by a lack of competent specialists.

6. System Maintenance and Reliability:

   Challenge: The CyberTwinAI systems are primarily based on AI algorithms and continual sources; because of this, system failure, inconsistent data, or biased algorithms might lead to wrong decisions or unplanned downtime.

7. Reliability and Resilience: CyberTwinAI Systems should be resilient and reliable; organizations should ensure this through tight quality controls, routine maintenance protocols, and redundancy to reduce such risk.
8. Compliance with Legal and Regulatory requirements:

   Challenge: CyberTwinAI can give way to legal and regulatory compliance problems when implemented into applications. This becomes more important in a high-regulated banking and healthcare industry. One needs to be careful while adhering to Safety, Ethics, Data protection laws, and regulations.
   Concern: This could also lead to fines against, litigation against, or even reputational damage to a company in case of non-compliance. Enterprises must ensure conformance of their deployments of CyberTwinAI to applicable laws and keep pace with relevant regulations.

6.11 Conclusion

Industry 4.0 future called CyberTwinAI. The impact it has on improving security and work output, particularly in the manufacturing and industrial industries demonstrates how innovation has adapted this industry to the fast-ever-changing environment. CyberTwinAI can help organizations When integrated and implemented with strategic planning, strict security measures, and ethical supervision, organizations may find it easier to not only survive but thrive even in the so-called digital age.

System integration, technology integration, and intelligence-driven management define the new industrial and manufacturing revolution, or what is known as Industry 4.0 has brought about. Despite the conflict between security and efficiency, both of the imperatives have emerged as fundamental as the environment changes incessantly. One radical solution to these issues is the introduction and use of CyberTwinAI, which is a new concept of integration of artificial intelligence with the digital twin. However, for organizations to thrive in the current world that is dominated by the use of the internet, some industrial operations must be efficient; this means cutting down on costs and maximizing production time as well as minimizing breakdowns. In terms of services, CyberTwinAI offers a broad perspective and assists business owners in anticipating issues, making decisions based on available information, and managing various physical systems to optimize their performance. Some scholars have proposed a more approach which they associated with Industry 4.0 From the details presented in the operational demands at the end of this year, technology's flexibility, expansiveness, and self-control functions are beneficial.

References

1. Sobb T, Turnbull B, Moustafa N (2020) Supply chain 4.0: a survey of cyber security challenges, solutions and future directions. Electronics 9(11):1864. https://doi.org/10.3390/electronics9111864
2. Hammad M, Jillani RM, Ullah S, Namoun A, Tufail A, Kim K-H, Shah H (2023) Security framework for network-based manufacturing systems with personalized customization: an industry 4.0 approach. Sensors 23(17):7555. https://doi.org/10.3390/s23177555
3. Javaid M, Haleem A, Suman R (2023) Digital twin applications toward industry 4.0: a review. Cogn Rob 3:71–92. ISSN 2667-2413. https://doi.org/10.1016/j.cogr.2023.04.003
4. Eckhart M, Ekelhart A (2019) Digital twins for cyber-physical systems security: state of the art and outlook. In: Biffl S, Eckhart M, Lüder A, Weippl E (eds) Security and quality in cyber-physical systems engineering. Springer, Cham. https://doi.org/10.1007/978-3-030-25312-7_14
5. Cali U et al (2023) Digital twins: shaping the future of energy systems and smart cities through cybersecurity, efficiency, and sustainability. In: 2023 International conference on future energy solutions (FES). Vaasa, Finland, pp 1–6. https://doi.org/10.1109/FES57669.2023.10182868
6. Pirbhulal S, Abie H, Shukla A (2022) Towards a novel framework for reinforcing cybersecurity using digital twins in IoT-based healthcare applications. In: 2022 IEEE 95th vehicular technology conference: (VTC2022-Spring). Helsinki, Finland, pp 1–5. https://doi.org/10.1109/VTC2022-Spring54318.2022.9860581
7. Wanasinghe TR et al (2020) Digital twin for the oil and gas industry: overview, research trends, opportunities, and challenges. IEEE Access 8:104175–104197. https://doi.org/10.1109/ACCESS.2020.2998723
8. Bamunuarachchi D, Banerjee A, Jayaraman PP, Georgakopoulos D (2020) Cyber twins supporting industry 4.0 application development. In: Proceedings of the 18th international conference on advances in mobile computing & multimedia, pp 64–73

Chapter 7
Digital Twin: Implementation

7.1 Introduction

The digital twins in the energy sector are enablers of the best asset performance, and the merger of green energy technologies ensures a dynamic but sustainable ecosystem. Healthcare practice areas include personalized care plans, surgical simulation, pharmaceutical manufacturing, and live patient monitoring to change the process of healthcare delivery. This chapter examines the digital twins application in transportation including predictive maintenance and fleet management ensuring efficiency and safety, also it analyses the issues regarding data privacy, model accuracy, ethical concerns and accessibility, making the security measures and open algorithms and equal technology access the most significant. At present, digital twins present huge benefits in medicine, though they support patient-centered care and treatment plans tailored for each individual patient [1].

The revolution of digital twin technology is one of the revolutions within the complex web of technological progress; it transforms theoretical models into tangible effects in different sectors. An analysis of the intricate processes that convert a digital twin from an idea into a real-life deployment is approached in the current chapter, comprising the technical fundamentals of digital twin implementations. Focusing on specific use cases in areas like construction, energy, healthcare, transportation, aerospace, agriculture, and retail, we start considering intricate aspects that push innovation and successful implementation [2].

We need to have the utmost familiarity with the peculiar problems in the field of digital twin as well as the new technologies and how to have them work together harmoniously from the stage of purpose to doing. To make digital twin projects work, people should cognize both businesses well enough, in a sense of how they are run, and the technical knowledge on how to turn abstract ideas to solutions, to be used.

In IoT devices, data analytics, and simulation approaches integration, it cleverly navigates the technical space of digital twin deployments in the energy industry,

© The Author(s), under exclusive license to Springer Nature Switzerland AG 2025

S. Gupta et al., *Digital Twins*, https://doi.org/10.1007/978-3-031-76564-3_7

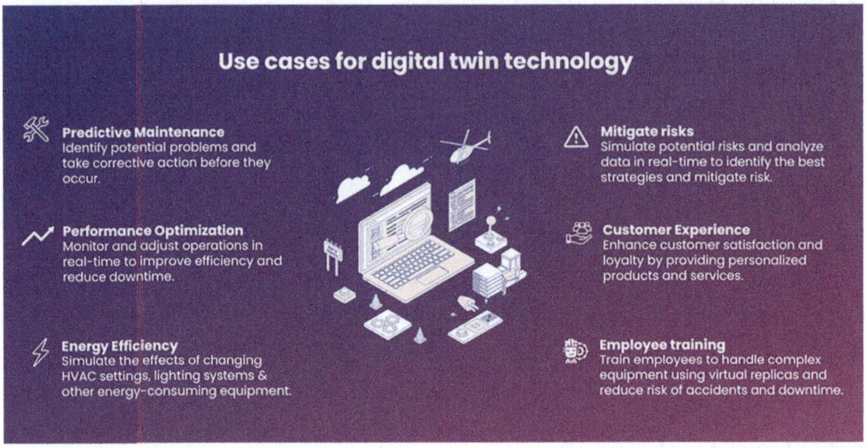

Fig. 7.1 Popular use cases of digital twins implementations Ref. [3]. (https://facilio.com/blog/digital-twins/)

optimal asset performance and the seamless integration of renewable energy sources are all illustrations of how digital twins have contributed to a dynamic and sustainable energy ecosystem [4].

In an inquiry into the marriage of electronic health records, data modelling and AI-based diagnoses, technological challenges of digital twin implementations in healthcare are revealed. The technological revolution has brought with it personalized treatment plans, surgical simulations, medication development, and real-time patient monitoring changing the way healthcare is given.

Digital twin instances that enhance predictive vehicle maintenance of the transportation sector technical, streamline fleet management, and predict traffic patterns. Digital twins are able to introduces the telematics, real time monitoring, and global positioning system technology innovations brought in the transport sector.

This chapter outlines the technical lessons learned from successfully wrapped digital twin projects. It provides an outlook, pondering what the digital twin technology is headed for and how it will transform the tech industry in the long run. A book is dedicated for professionals, researchers and fans to gain clarity about workable mechanisms of digital twins in different spheres (Fig. 7.1).

7.2 Use Case 1: Electrical Vehicle Battery Digital Twin

The battery DT (Digital Twin) technology is advantageous in different terms as outlined in the following discussion. Battery DT technology has many benefits, some of which are listed below: Battery DT technology has many benefits. It eases the optimization, operation, and maintenance of the battery design. Batteries are also designed to be economical. It reduces the effect of climate change by minimizing transportation related emissions and making renewable energy more affordable. Promoting renewable energy sources and reducing carbon emissions makes one part of the UN sustainable development goals done which are environmental, society, and government. As it encompasses a detailed comprehension of the battery behavior and performance during its life cycle, it helps in achieving policy targets associated with energy storage, emissions reduction, and sustainability. Standards and Regulation accelerate the attempts to control and standardize battery performance and safety. The 2020 number of publications on battery DT research has tripled which is a sign of high interest in this topic. Most of these studies have used the battery DT to get the SoX (State -of – X) estimate, design optimization, production, and monitoring. Lately, battery DTs have become popular for second life applications and predictive maintenance. Energy optimization, optimized battery charging, life management, and battery passport are also considered as applications of the battery DT framework. The achieved results of 60% maintenance cost reduction, 15% lifetime improvement using optimal charging methods, and DT-based SoC estimation MAE (Mean Absolute Error) of 0.14% is encouraging for battery DT application. Software, hardware and IoT are three classifications of cross-discipline battery DT elements and criteria. Each group is reviewed, and then technical details are addressed. We assess sensor networks that contain VIT (Voltage, Current, Temperature) [5].

The concern for the environment and attempts to decrease the use of fossil fuel have pushed the sales of the EV. Considering battery-operated electric cars, searching for improvements regarding performance, energy efficacy, and duration still remains to be done. This chapter investigates the implementation of digital twin as a betterment approach of EV batteries. Digital twins of electric battery living data models of electric car batteries. The continuous monitoring of sensor data on the battery health, performance, and operation enables digital twins to detect problems at the early stage hence, predictive maintenance. With such an active approach, the battery life is prolonged, and related problems are avoided [6].

Test operating states, charging regimes and thermal control strategies to which digital twins also enable engineers and scientists to virtualize are used in the electronic equivalent of an experiment to tune energy storage, rate of charge, and battery performance. Nevertheless, digital twin battery simulation is challenging. Modern modelling similarly should render an adequate model of electrochemical behavior, thermal properties, and aging. The data of physical battery that is used for the update of the digital twins should be safe and trustworthy as such to ensure credibility and protection from unauthorized access (Fig. 7.2).

(i) Detailed modelling

Detailed digital twins give information about complicated electrochemical processes, thermal, and aging of EV batteries.

Actual Case: Electric automobile li-ion battery cell modeling. Modeling of battery cell electrochemical processes and heat generation are done in COMSOL Multiphysics and ANSYS Fluent, respectively.

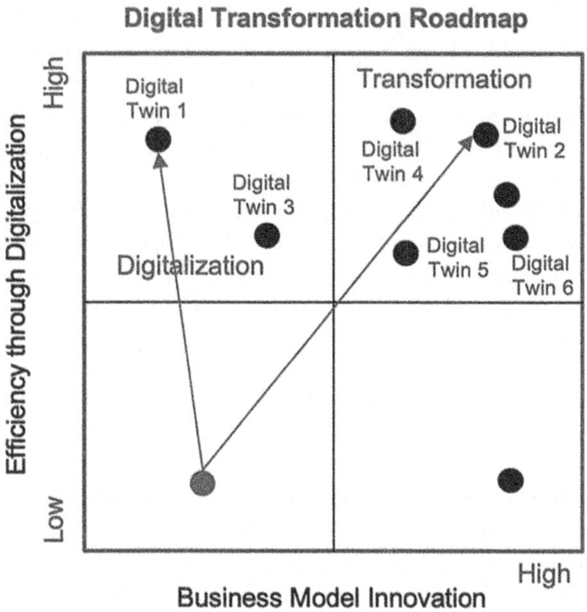

Fig. 7.2 Implementation of digital twins for EV batteries Ref. [2]

(ii) Battery Health Monitoring and Forecasting:

Digital twins for real-time battery sensor data alert proactive maintenance and health monitoring receive failures and performance degradation.

The digital twin technology enables Tesla to continually assess the condition of the battery. Continuous observation of the state of charge of the battery makes the company notice the problem in time and then correct the problem in order to protect performance and lifetime.

(iii) Improving Battery Performance: The digital twins replicate the operating, charging and thermic management of EV battery to enhance the effectiveness. Through digital twins, GM (General Motors) enhances storage and charging of electric vehicle batteries. Testing and management in virtual will help GM (General Motors) improve energy efficiency and performance of EV batteries [6].

(iv) High-Tech Control System Integration. Technical Details: Digital twins with powerful control algorithms and energy management systems enable immediate battery power optimization.

Actual Case: The new energy management algorithms and digital twin technologies improve the battery consumption of BMW i3 car. The live data of the digital twin makes the power distribution of the vehicle energy management system to optimize for efficiency and range.

(v) Machine Learning Prediction Improvement: by the failure of such companies that were meant to provide ISO-standards-oriented management systems. Machine Learning Prediction Improvement:

Technical Info: The pattern evaluation through the machine learning algorithms and digital twins of big battery data for the betterment of predictive maintenance and performance.
Actual Case: Nissan is assessing digital twins and machine learning in the context of predictive maintenance of electric vehicle batteries. Data on consumption and efficiency permits technology to expect degradation and to recommend maintenance.
Illustrations and cases indicate how digital twin technology improves the function, reliability, and effectiveness of an electric car battery system. Evolution of electric cars will be dependent on modeling, forecasting capabilities and integration side of the evolution activities [7].

Implementation of EV battery digital twins require data gathering, model creation, integration with live data sources, predictive Here are EV battery digital twin implementation and application scenarios: Implementation of EV battery digital twins require data gathering, model creation, integration with live data sources, predictive Here are EV battery digital twin implementation and application scenarios [8]:

(i) Data Collection: A Battery monitoring system is installed by an electric vehicle manufacturer which monitors temperature, voltage, current and charge; these are state-of-the-art devices. Central databases store and analyze the data.
(ii) Model Development: – The use case: An accurate digital twin model of the EV battery behavior is developed by the builder using the information. This model, for detailed simulations and analysis, takes into account battery electrochemical, thermal and ageing characteristics. The digital twin is linked to real time EV fleet data systems for the update of virtual model with current operational data. From this connection, the digital twin replicates the batteries performance.

    The structures of the model are created in such a way that some parameters have an effect on the behavior of the model if specific inputs are provided. In other words, it is possible that certain parameters are never discovered through the process of calibration. With regard to these particular instances, it has been established that noise filtering techniques and the design of studies are beneficial. An additional obstacle is the time-varying behavior of the model, which leads to differences in the parameters of the model under different situations. These conditions include different temperatures, standard operating conditions (SoC), C-rates, and degradation levels. In order to overcome this obstacle, a number of different real-time model parameterization strategies have been suggested. The methodologies that are most widely used are those that are founded on recursive filtering techniques, such as the least-squares family. More recently, the application of machine learning for the parametrization of battery models has also been investigated [9] (Table 7.1).
(iii) Predictive Analysis and Optimization: The digital twin forecasts battery problems and deterioration with live data. The digital twin helps to prevent battery

Table 7.1 Comparison of different battery models used in digital twins Ref. [5]

Equivalent circuit models	Pros	Cons
Model	Widely used for SoX estimation, a nice trade-off between accuracy and complexity, error lower than 2% achievable	Parameterization requires special types of tests at different temperatures and SOC conditions, poor interpretation of battery physics
Electrochemical model	High accuracy, ability to track the aging of battery components, e.g., the thickness of the solid electrolyte interphase, models the physics of battery	Prior knowledge of battery required, Manufacturing and construction data needed, time-consuming training process, high computational burden
Data-driven	Does not require prior knowledge about the cell, strong mapping potential	Collecting the training datasets in the laboratory is very time-consuming and costly
Empirical -f(SoC.I.T)	Simple representation, minimum complexity	Low accuracy

issues by alerting the maintenance team to the unequal temperature distribution and capacity reduction.

(iv) Performance Optimization and Control: Digital twins improve the performance of the EV battery. The digital twin, by modeling several recharging strategies and operation scenarios, can provide the optimal charging profiles and energy management that increases battery life and vehicle range.

(v) Integration with Energy Management Systems: Application: The digital twin maintains efficient EV battery utilization on the fly. The digital twin's data, obtained from different sensors in real time by the energy management system allows modifying power distribution, regenerative braking, and other operational aspects, and hence enhances energy economy and vehicle performance [10].

(vi) Continuous improvement, iteration: New data and information that digital twin brings lead to the refining of the model that is aimed at better forecast and function. The actual battery performance and maintenance data improves the digital twin model. The methodologies and use cases can be utilized by the organizations to improve electric vehicle battery performance, reliability, and efficiency. Such an approach improves predictive maintenance, energy management, and vehicle performance, thus elevating the levels of electric car adoption.

Utilizing the components of electrical circuit, equivalent circuit models are efficient and capable of displaying the activity of the battery. Such model networks are not able to precisely recap the physiochemical characteristics of the cell, which are why the effects of the device on the cells are not sufficiently accurate. Moreover, it turned out that they are great for specific reaches, like decisions of what-if scenarios and identifying malfunctions. A model based internal short-circuit detection technique has been developed which has been engineered to accommodate faults and noises from the aging equipment. Thus, based on Thevenin models, an analogous analog circuit has been thrown in to the design. Since conditions in surrounding differ, including that of different times, various battery models have been developed in a random manner. These models have also been introduced in the prediction of battery ageing behaviour. Nonetheless, the present project does not have a full assessment of battery models. Over the last years, there have come to be lots of very good papers that have been published specifically on the modelling subject [11].

Effecting this in reality the DT may be a system including a cluster of the models, and then the connection with subsystems, smaller systems and processes. Cell models simulate the basic constituent, while DT is the layering or joining of these models in theory. It can be done the way where cell and other subsystem models are unified, for instance, to construct the module and pack models – where the pack model may also contain the model of the BMS/TMS- applying a one-level approach. The illustration presented in Fig. 7.3, Fig. 7.3, conveys the concept. In that aspect, the strategy to create an outstanding model as well as the improvement toolkit are

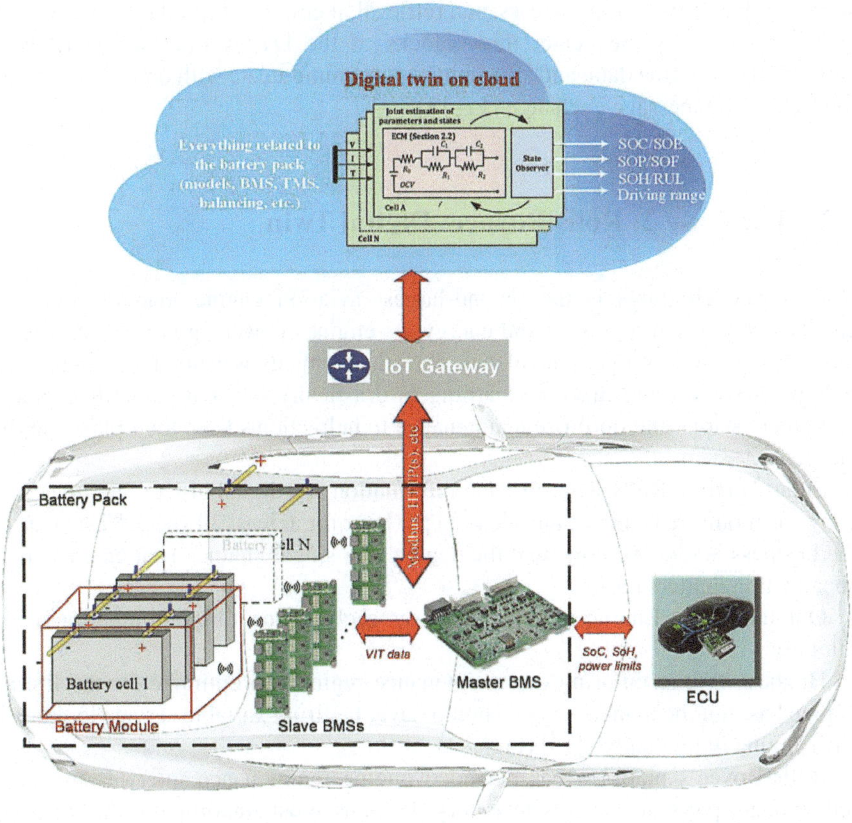

Fig. 7.3 DT battery models and physical battery architecture with BMS/TMS Ref. [5]

noteworthy features to be mentioned. DT boxing model utilizes all the criteria of multicellular models, consequently integrated cell models in it, modeling scenarios in to what-if, and updating model by computing are taken into consideration [12].

Digital Twin Layout The utilization of platform-integrable services is a hallmark of DT frameworks. State estimation (also called inverse modelling or sensor fusion), predictive maintenance, problem detection, decision-making assistance, and self-adaptation are the most significant services for DT. Layers allow us to understand the DT's architecture: the bottom layer acts as an Internet of Things (IoT) framework, allowing the DT and the PT to share data and connect to one another, and the top layer offers more advanced services. We first used the term "3D DT" to describe DTs using just the bottom layer to describe these layers. Adding the data and service dimensions to the 3D DT, the concept was referred to as 5D DT. In addition to a service system for deploying data-driven services, the 5D DT offers a central repository that is linked to the physical, virtual, and service spaces. Following the implementation of 3D DT in the industry, the necessity for 5D DT emerged as a result of the new services that provide higher-value functionality, Fig. 7.3 shows the fundamental design of a DT. For inter-service communication, it employs a RabbitMQ server, and for data storage, access, and retrieval, it employs InfluxDB. The services are represented by the center three blocks of the DT. A time series database (InfluxDB) stores the data, and all components communicate with one other through the RabbitMQ message exchange [11–13].

7.3 Use Case 2: Rolls Royace: Digital Twin

Rolls-Royce Holdings is the second-largest aviation engine manufacturer and engages in marine propulsion and energy. Its engines power fighter jets, corporate jets, and over 50% of long-haul planes. Now the company is using digital twin technology, analytics, and machine learning to drastically lower its aircraft engines' carbon emissions and optimize maintenance to help clients keep their planes in the air longer.

Stuart Hughes, Rolls-Royce's chief information and digital officer, says the company has monitored engines and charged per hour for at least 20 years. "That part of the business is old. We now treat the engine as a single engine. That engine's personalization matters more."

Our Intelligent Engine platform analyses each engine's flight, conditions, and pilot use.

Hughes: "We're tailoring our maintenance regimes to optimize for the life an engine has, not the manual says it should have. It's truly variable service looking at each engine individually" [14] (Fig. 7.4).

Rolls-Royce's platform has extended engine maintenance intervals by 50%, reducing its parts and spares inventory. Perhaps most importantly, the business claims it has saved 22 million tons of carbon by improving engine efficiency. Even

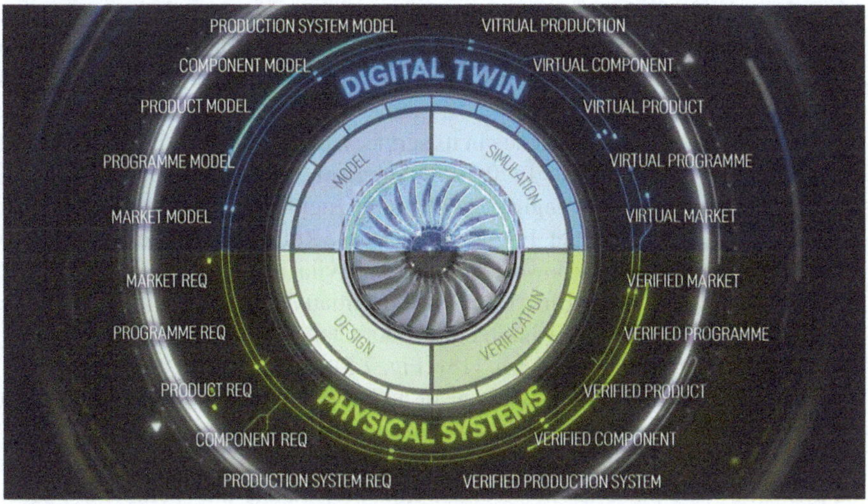

Fig. 7.4 The Rolls-Royce Digital 'O'—a foundational strategy for digital twinning Ref. [14]

Rolls-Royce is employing AI to learn how to handle metal refuse and garbage from retired parts. Hughes says, "Since 2014, it helped one of airlines avoid 85 million kilograms of fuel and over 200 million kilograms of carbon dioxide. It is done by taking data on how the pilot flies the plane, operates it, and funds it operationally. Collected data and analytic insights helped to improve the judgements. In places where change was difficult, new policies and procedures were established" [15].

Rolls-Royce uses two-way, real-time engine data and batch information after jets land to power their Intelligent Engine platform. The data is fed into a Microsoft Azure data lake and then into a Databricks "lakehouse" for machine learning and AI. Databricks calls its open architecture a data lake-data warehouse hybrid. Each flight generates around half a gigabit of data from real-time and post-landing data. The company's "Engine Condition Monitoring" service analyses real-time data for engine performance irregularities to predict repairs. The analytics can predict in-flight whether a thorough landing check is needed, helping the airline prepare and minimise travel disruptions. More sophisticated prediction modelling can employ other data [16].

Top of the list reasons of aircraft on ground (AOG) is unscheduled maintenance. The U.S. airlines in 2019, the last year with continuous airline operations, reported domestic tarmac delays of more than 3 h 302 times according to the data released by the U.S. Bureau of Transportation, which was an increase from 202 in 2018 and 193 in 2017. The aim of Rolls-primary Royce is to make all aircraft powered by its systems operate on time, continuously, and with almost 100% availability [17]. These benefits of the aircraft's owners, Rolls-Royce, the original equipment manufacturer, and the passengers. It is at this point that the incentives for predictive maintenance for all stakeholders operating in a specific route converge. Blue Data Thread is a significant contribution to Rolls-approach Royce for eliminating unforeseen downtime. A jet engine is almost incomprehensibly high-tech and complicated stuff, but Rolls-Royce ensure that their A330 engines can complete 50 roundtrips to the moon and back between overhauls by being in tune with their performance and maintenance requirements. Aviation has moved to predictive maintenance over time. Rolls-Royce uses data science in its services and solutions to increase aircraft availability for clients. In the 1990s, IFS Maintenix specialists participated in a U.S. Navy study to predict engine failure using engine component data analysis. Ward added, "The Intelligent Engine emphasizes Rolls-Royce's commitment to electrification and other new power sources that are vital to air quality. IFS Maintenix is part of the Rolls-Royce Blue Data Thread initiative, which enables predictive maintenance for all engine life-limited parts. Assistance is provided from manufacturing to operation and maintenance [18] (Fig. 7.5).

Since 2014, Rolls Royce assisted one of its airlines in avoiding the consumption of 85 million kilos of gasoline and more than 200 million kilograms of carbon dioxide, as stated by Hughes. Through the collection of data on how the pilot is flying the plane, how the plane is operated, and how they handle the operational budget around that, we were able to accomplish this. Using Digital twins, it could be possible to provide them with data and insights that assisted them in improved decision-making. The areas in which they believed there were obstacles to change were the ones in which Digital Twin assisted them in developing new policies and procedures.

Rolls-Royce: The Power of Digital Twins Unmasked Embracing the digital revolution was a major leap that Rolls-Royce took from its tradition that was deeply rooted in engineering innovation and precision. By means of the integration of digital twins, Rolls-Royce has revolutionized its approach to conceptualizing, designing, manufacturing, and maintaining its renowned engines, therefore leading the

Fig. 7.5 Rolls Royce digital twin Ref. [19]

company to a new era of perfection. This chapter goes deep into the details of Rolls-Royce's digital twin journey, outlining how the digital twins have evolved into the maestros who are at the helm of the innovation symphony. The journey of Rolls-Royce into the domain of digital twins started as a desire for deeper understanding and management of their intricate engineering masterpieces. It became apparent that the traditional philosophy towards design and maintenance was inadequate in the context of the modern requirements of aerospace and marine sectors. The company, in turn, adopted digital twins, which were virtual copies that looked and acted like the original objects in every respect [18].

Digital Precision Engineering Rolls-Royce digital twins are not merely simulations, they are complex, dynamic replicas that capture lifetime of their products. From conceptualization to design, manufacture, and ongoing maintenance, these digital twins reflect their physical counterparts through real-time, data-driven simulation.

It is in the synergy of simulation and reality where the true force of digital twins comes into play. Rolls-Royce engineers exploit cutting-edge simulation methods to try out design changes, test hypotheses and to winnow engineering solutions. The knowledge obtained from these simulations drives the design of the solid product and thus a 'feedback loop' of improvement is created. Relying on this virtual/tangible play will allow Rolls-Royce to continually break new frontiers in innovation and at the same time keep steadfast commitment to the key points of safety and dependability. The digital twins are a connecting link between theory and practice, giving the chance to engineers to try out with new ideas in the virtual environment before they are finally implemented in the real world.

Rolls-Royce has been very careful when it comes to sharing the intricate technical details about its digital twin activities with the public. We are delivering a brief outline of the main technological components that are commonly found in digital twin deployment in Rolls-Royce basing on the information that has been provided (Fig. 7.6).

(i) Data Acquisition and Integration: Collecting and integrating diverse data sources is the first stage of implementing Rolls-Royce's digital twin. There are pieces that compose of this that are, real time data from engine sensors, performance records, maintenance logs, and environmental variables. In order to gather data, complex sensor networks are installed on automobile engines and other vital components. These networks are monitoring quite a few of parameters including the fuel amount, temperature, pressure, vibration and so on.

Sensor technology: Rolls-Royce uses a diverse range of sensors that collect real-time data from critical components such as engines. Some of the data these sensors monitor include vibration, temperature, pressure, fuel flow rate, exhaust gas composition and many others. Sensor technology used includes the following:

Melting pot is not only a common identity but also a common lifestyle now.

Fig. 7.6 Digital twin of aeroplane engine Ref. [20]

Strain gauges are used to measure the level of mechanical stress and deformation in engine components. *Thermocouples*, infrared sensors and resistance temperature detectors (RTDs) are the instruments used to monitor temperature gradients in numerous engine components.

Accelerometers are able to sense the vibrations and oscillations in the components of an engine and therefore, they can give a lot of information about the engine's performance and structure.

Flow meters: The instruments measure the velocity at which the engine fluids of several types such as petrol, oil and coolant pass through the engine.

To transfer engine sensor readings to the digital twin platform, data acquisition devices are utilised. These devices process and transmit the data. One component that is included in these systems is the data acquisition unit, also known as the DAU: In order to digitize analogue signals and transmit the data to systems on the ground, DAUs are designed to be deployed onboard ships or aeroplanes. They connect to sensors to accomplish this job. *Telemetry* systems enable data to be wirelessly delivered from remote locations to centralised monitoring centres. This is accomplished through the utilisation of cellular or satellite information networks. Embedded systems are microscopic computer units that are built into various engine components to process and analyse data onboard (Fig. 7.7).

Modelling and simulation: Rolls-Royce's digital twins are developed using state-of-art modelling and simulation tools to create digital replicas of real assets. A hybrid approach of physics-based modelling techniques and high-fidelity CAD data is used for creating these models. Numerous simulation approaches, for instance finite element analysis (FEA) and computational fluid dynamics (CFD), are used to precisely simulate the behavior and performance of engine parts in

Fig. 7.7 Digital twin development—Rolls Royce Ref. [21]

different operating conditions. Multi-bodied Dynamics (MBD) software is used to simulate the dynamics and interactions of assemblies of connected mechanical parts. This contains shaft systems as well as other systems such as rotor assemblies [21].

Rolls-Royce utilizes a digital twin platform composed of visualization tools, analytics, and real-time monitoring to do data analysis and give valuable insights. An essential component of this platform is comprised of the following:

Data processing pipelines and preprocessors together with raw sensor data will gather, clean, and prepare the data for analysis. There are mountains of data that can be through algorithms which are driven by machine learning and artificial intelligence. They can be used to search for patterns, outliers, and upcoming failures of equipment.

Visualization and dashboard tools include the following: Provide simple ways for operators and engineers to see diagnostic warnings, key performance indicators, and statistics related to engine performance.

Edge Computing and Cloud Integration Rolls-Royce digital twin deployment is backed by a hybrid of cloud-based infrastructure and edge computing which brings us to our fourth point that is computing integration with cloud. Edge computing is a solution that allows for processing and analysis to be done near the source of data with a consequent reduction in the volume of bandwidth needed and a shortening of reaction times. Cloud integration makes scalability and flexibility possible as well as centralized storage of large volumes of data that is produced by digital twins for a number of assets and from different locations.

Cybersecurity and Data Privacy Due to the criticality of the aerospace and marine industries, Rolls-Royce highlights cybersecurity and data privacy in its digital twin implementation. Powerful security tools are put in place to guard the sensitive data and guarantee the integrity and confidentiality of information shared between the physical assets and digital twins. Conformance to industry regulations and standards i.e. GDPR (General Data Protection Regulation) and ISO 27001 is also a priority.

Rolls-Royce employs a range of cybersecurity technologies and best practices to safeguard sensitive data and protect against cyber threats:

Encryption and Secure Communication Protocols: Maintain data integrity and confidentiality when transmitting data between sensors, edge devices, and cloud-based systems.

Access Control Mechanisms: Limit unauthorized access to the digital twin platforms and provide role-based permissions for users and administrators.

Security Monitoring and Incident Response: Regularly check the system log and the network traffic for any possible suspicious phenomenon and implement the standard operating procedures for the detection, reaction, and recovery.

Collaborative Ecosystems and Integration with Partners: The client is also considering the option of visiting one of these countries in order to have a face-to-face meeting with potential customers.

Collaboration of Rolls-Royce is often made with the network of partners, aircraft manufacturers, airlines, and maritime operators to improve the functionality of their digital twin implementations. Integration with third-party systems and platforms is achieved via standard interfaces and APIs (Application Programming Interfaces), enabling smooth data exchange and interoperability within the system.

Although these technical aspects give some sort of "general guidelines" for the Rolls-Royce digital twin implementation, proprietary technologies, algorithms, and architecture details might not be disclosed to the public. To obtain the most recent technical updates and innovations, ones should consult the official publications, technical whitepapers, and presentations at industry conferences and events by Rolls-Royce [21].

7.4 Use Case 3. Mayo Clinic: Digital Twin

The computation of diagnostic metrics in the fields of coronary artery disease, aortic aneurysm, aortic dissection, valve prosthesis, and stent design. Mayo Clinic has been actively researching digital twin technology for many years. Ognjen Gajic, M.D. and his colleagues are researching the possible utilisation of a digital twin patient model to forecast the therapeutic response of patients within the initial 24 h of sepsis. Their research proved the viability of developing and prospectively testing a causal artificial intelligence model to predict therapy response in the early stages of critical illness. The ICU is an attractive setting for the development and testing of digital twin patient models due to the presence of both qualitative and quantitative data, as well as a reasonably quick turnaround time (Fig. 7.8).

The term "digital twin" might make you think of people being copied in labs, like in the movie The Matrix, if you're into sci-fi stuff. But it's cooler than that. A digital twin is like a perfect virtual copy of a real thing. Lots of people in the healthcare industry are starting to use this technology. Cardiology and critical care medicine are leading disciplines in the field of digital medicine. For patients with arrhythmias, a digital blueprint can offer valuable information about the underlying mechanical discoordination in the heart and enhance the effectiveness of cardiac resynchronization therapy (CRT).

In order to help physicians better understand the connections between genetic variations and specific illnesses, digital twin modelling is currently being used to represent an individual's genome. Advance of Technologies provides a platform and interface to access genomic data that is essential for personal health records. The interface shows a graphical representation of an individual's entire gene pool as well as specific gene variations associated with recognized medical conditions. These variations are correlated with different body systems, value sets, and clinical code sets, enabling tracking over time [22] (Figs. 7.9 and 7.10).

The Digital Twin Revolution: How Mayo Clinic is transforming patient care the Mayo Clinic, famous for its innovative medical developments, introduces the digital twin. Its embedding in the health care system opens up a lot of possibilities for personalized medical care and improved patient outcomes. Digital twin technology in Mayo Clinic is approached through its methods, potential gains, and ethical concerns.

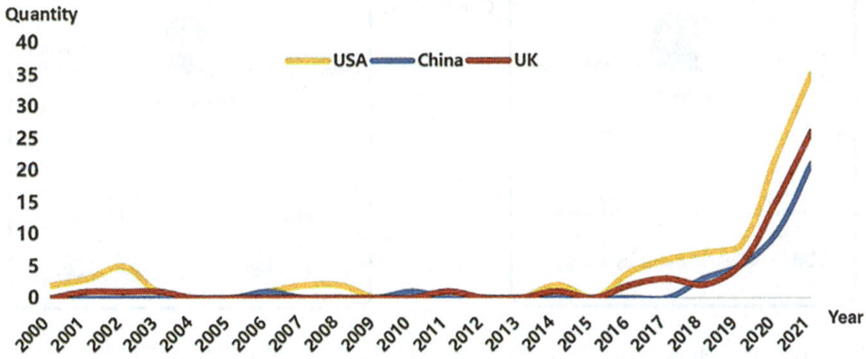

Fig. 7.8 US, UK, and Chinese medical digital twin technology investigations from the past 20 years. Ref. [22]

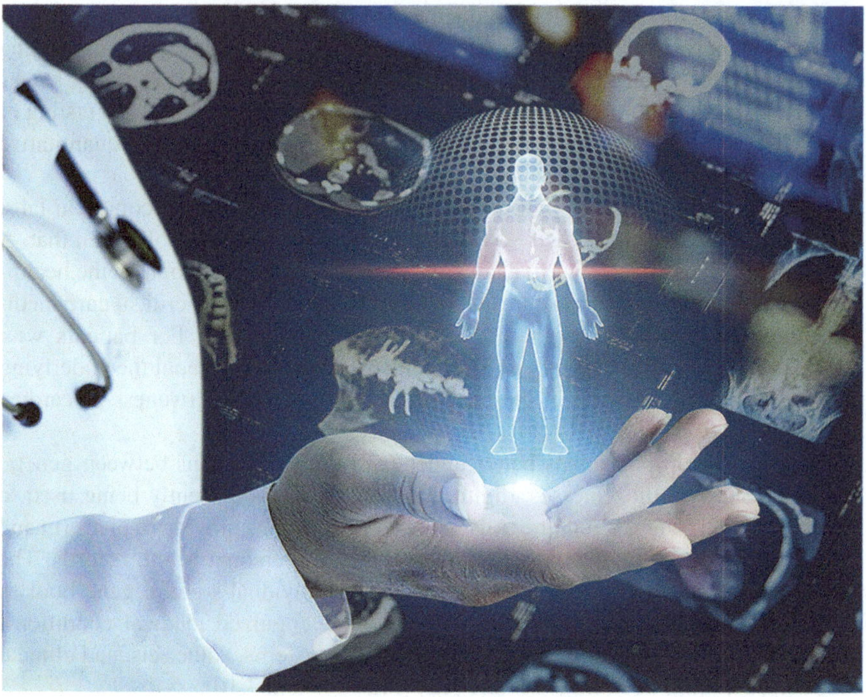

Fig. 7.9 A digital twin of a medical patient Ref. [22]

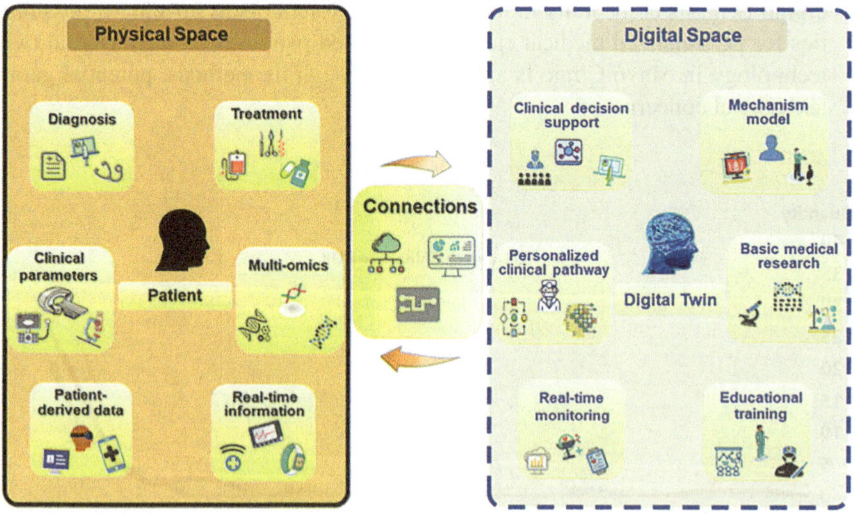

Fig. 7.10 Application and prospects of digital twin technology in the medical field. Ref. [23]

Developing Virtual Patient: Creating a digital twin of every patient is an essential part of this project. This virtual edition integrates data from many diverse sources, for instance,

Electronic health records (EHRs): A centralized data source that has historical medical records, diagnostics, medications prescribed, and treatment histories.

Genomic data: Genetic information analysis of an individual provides personalized understanding of disease susceptibility and potential response to treatment.

Wearable devices and sensors: Active monitoring of vital signs, activity levels, and patterns of sleep provides instantaneous information about the patient's health status.

Medical imaging: Detailed information about anatomy from the scans and the imaging aids in understanding the disease progression.

These data points are used in a computational model to replicate the unique physiological attributes of the patient and how the patient responds to different stimuli. This facilitates:

Predictive modeling: The digital twin can forecast possible reactions to various kinds of treatments, drugs or lifestyle changes and thus enables doctors to make decisions based on information.

Virtual trials: Performing experiments of novel drugs or treatment protocols on the digital twin helps in prediction of possible hazards and optimization of tactics, prior to their application on real patients.

Personalized medicine: With digital profiles of each person deemed unique, treatment plans can be personalized about the individual, and thus doctors are able to go beyond traditional and uniform models.

Using digital twin technology capabilities, Mayo Clinic targets to transform healthcare delivery, promoting personalized care and better patient outcomes [24].

Challenges and Ethical Considerations Despite its promise, implementing digital twins in healthcare faces several challenges:

- Data privacy and security: Patient data protection is pivotal that needs robust security measures and defined ethical guidelines.
- Model accuracy and validation: The digital twin simulations being very accurate and reliable is a condition for the clinical utility of the digital twin.
- Ethical considerations also play a vital role: The organization stipulates that no one should talk on their cell phone while in a meeting.
- Informed consent: Patients should be told clearly how their data is being used and the potential consequences of the digital twins.
- Transparency and explainability: Transparent models and simple algorithms that are understandable to humans foster trust and accountability.
- Accessibility and equity: Equal access to this technology to all patients is imperative regardless of their socioeconomic status.
- Bias and fairness: The models must be carefully designed in such a way that they don't perpetuate biases in the healthcare data.

- The Future of Digital Twins in Healthcare: The digital twins being developed at Mayo Clinic is a huge leap in personalized medicine, however, though there still persist some challenges the possible benefits are colossal. Through technology development and solution of ethical issues, digital twins can be a way for healthcare systems to become a proper patient-centered care.

Note: This article is derived from publicly accessible sources and does not claim to present the implementation particulars of Mayo Clinic's digital twin project. For correct and up-to-date information official sources should be checked [24, 25].

Patient-Specific Digital Twin Development and Implementation A deep discussion on the methodologies employed in the creation, maintenance, and utilization of patient-specific predictive digital twins is needed prior to implementation. It begins by providing an overview of the constituent elements of the predictive digital twin, contextualizing them within relevant fields. Following this, needed to be delves into the mechanistic model, examines treatment control parameters, and outlines the process of generating observational data for simulated patients. Additionally, Bayesian model calibration may require for the propagation of uncertainty for patient-specific prognoses, and the implementation of multi-objective risk-based optimization under uncertainty. Finally, adhering the utilization of survival analysis methods for assessing the effectiveness of optimized treatment plans.

Current Applications of Digital Twins in Medicine Personalisation and patient-centric modelling are key to the realisation of precision medicine. With the fast expansion and advancement of several activation technologies, the use of the DT in the medical sphere has immense potential. As the cost of IoT devices falls and they become easier to integrate, the actual network connectivity increases. DT healthcare was offered as a new medical simulation method for providing resilient, precise, and effective medical services by combining technology with interdisciplinary, multi-physics, and multiscale models. We can employ AI and other cutting-edge technology to properly determine the origin of a patient's ailment, clarify the therapy aim, and provide personalised and precise care. Several preliminary investigations have created the groundwork for further use of the DT in medicine [26].

Orthopaedics (n.d.) With regard to dynamic observation, the first orthopaedic digital twin research was conducted with the intention of addressing the deficiencies that exist in the existing methods of biomechanic research. With the help of the tailored data of a specific experimenter's lumbar spine bones, researchers were able to construct the shape-performance integrated DT body. This allowed them to anticipate the biomechanical properties of the actual lumbar spine under a variety of human postures. An individual's body motion, stance, and spatial position were recorded in real time by a technology called human motion capture. The lumbar position of the human body was controlled to a certain extent by the wearable virtue actual reality gadget and a small amount of sensor data. In order to recreate a wide range of human activities and postures, the DT body of the lumbar spine was developed with the assistance of the inverse kinematics technology and the finite element method. The AI model performs calculations in real time, which are then visualised and incorporated into the virtual lumbar spine. These calculations are based on the data that is collected from the posture. The analysis of lumbar spine biomechanical features and the prediction of those characteristics allowed for the successful completion of real-time monitoring and prediction efforts. For the purpose of capturing the real-time biomechanical performance of the lumbar spine, the Unity3D software was utilised in the development of a 3D virtual reality system. This technology has the potential to alter the ways in which patients who have suffered spinal injuries have been scheduled to be treated [27–29].

7.5 Conclusion

This chapter speaks of digital twin adoption across industries and its disruptive force. Energy efficiency and patient care are also challenges that are addressed with digital twins. Detailed use cases analysis of Electric Vehicle Battery Digital Twin, Rolls-Royce Aircraft Engine Digital Twin, and Mayo Clinic Healthcare Digital Twin show the practical advantages and applications of digital twins [26].

However, digital twins are a land full of opportunities and challenges. To use this technology in a reasonable and equitable way, the issues of data privacy, model consistency, ethics, and accessibility must be addressed. Digital twin systems necessitate strong security, open algorithms, and equitable technology access to remove threats, and build trust.

Even though digital twins in health are restricted in number, they have great prospects. The power of digital twins lies in their ability to transform healthcare delivery through personalized care, predictive modeling and virtual trials, all of which improve patient outcomes and quality.

Digital twin technology is a migrating technology and hence the stakeholders must migrate with it, collaborate, innovate, solve problems. Combinations of digitals can bring new success to the industries, stimulate creative development and increase the sustainability and efficiency of the industries.

This chapter deals with the practical applications, challenges, and opportunities of the digital twin technology. Digital twins technical fundamentals, real-world scenarios of application, and moral issues are utilized by professionals, researchers, and hobbyists to understand, discuss and use this innovative technology productively in many industries.

References

1. Chaudhary G, Khari M, Elhoseny M (2021) Digital twin technology. CRC Press, Boca Raton
2. Nath SV, Van Schalkwyk P, Isaacs D (2021) Building industrial digital twins. Packt Publishing
3. Rao M (2023) How to Develop and Implement Digital Twin Technology? Facilio Blog. [Online]. https://facilio.com/blog/digital-twins/

4. He X, Ai Q, Pan B, Tang L, Qiu R (2022) Spatial-temporal data analysis of digital twin. Digital Twin 2:7
5. Naseri F, Gil S, Barbu C, Cetkin E, Yarimca G, Jensen AC, Larsen PG, Gomes C (2023) Digital twin of electric vehicle battery systems: comprehensive review of the use cases, requirements, and platforms. Renew Sust Energ Rev 179:113280
6. Wang W, Wang J, Tian J, Lu J, Xiong R (2021) Application of digital twin in smart battery management systems. Chin J Mech Eng 34(1)
7. Bhatti G, Mohan H, Raja Singh R (2021) Towards the future of smart electric vehicles: digital twin technology. Renew Sust Energ Rev 141:110801
8. Gururaj HC, Hegde V Digital twinning of the battery systems—a review. Lecture notes in electrical engineering, 21-Aug-2021. [Online]. https://doi.org/10.1007/978-981-16-1642-6_11
9. Wu B, Widanage WD, Yang S, Liu X (2020) Battery digital twins: perspectives on the fusion of models, data and artificial intelligence for smart battery management systems. Energy and AI 1:100016
10. Saeed Z, Mancini F, Glusac T, Izadpanahi P (2022) Future city, digital twinning and the urban realm: a systematic literature review. Buildings 12(5):685
11. Kharlamova N, Hashemi S (2023) Evaluating machine-learning-based methods for modeling a digital twin of battery systems providing frequency regulation. IEEE Syst J 17(2):2698–2708
12. Naseri F, Schaltz E, Stroe D-I, Gismero A, Farjah E (2022) An enhanced equivalent circuit model with real-time parameter identification for battery state-of-charge estimation. IEEE Trans Ind Electron 69(4):3743–3751
13. Rjabtšikov V, Rassõlkin A, Kudelina K, Kallaste A, Vaimann T (2023) Review of electric vehicle testing procedures for digital twin development: a comprehensive analysis. Energies 16(19):6952
14. Digital FIRST. [Online]. https://www.rolls-royce.com/products-and-services/defence/digital-innovation.aspx
15. Ansys collaborates with Rolls-Royce and Intel on new turbine tech simulation. Military Aerospace, 14 Jun 2023. [Online]. https://www.militaryaerospace.com/commercial-aerospace/article/14295179/ansys-collaborates-with-rollsroyce-and-intel-on-new-turbine-tech-simulation
16. Elliott J How Rolls-Royce and IFS are creating the 'IntelligentEngine' of the future. IFS Blog, 09 Aug 2021. [Online]. https://blog.ifs.com/2021/08/how-rolls-royce-and-ifs-are-creating-the-intelligentengine-of-the-future/
17. Geo Rolls-Royce turns to digital twins to improve jet engine efficiency. Geonation®, 11-Jun-2021. [Online]. https://geonation.tech/rolls-royce-turns-to-digital-twins-to-improve-jet-engine-efficiency/
18. Du Preez D How Rolls-Royce is improving engine sustainability with real-time data and digital twins. diginomica, 22 Nov 2022. [Online]. https://diginomica.com/how-rolls-royce-improving-engine-sustainability-real-time-data-and-digital-twins
19. Evolution, not revolution. [Online]. https://www.rolls-royce.com/country-sites/india/discover/2020/evolution-not-revolution.aspx
20. Rolls-Royce outlines a vision for the future with its digitally driven IntelligentEngine, 06 Feb 2018. [Online]. https://corporatetravelcommunity.com/analysis/rolls-royce-outlines-a-vision-for-the-future-with-its-digitally-driven-intelligentengine-583120
21. Digital Twin. [Online]. https://www.rolls-royce.com/innovation/digital/digital-twin.aspx
22. Personal digital twins come to healthcare.—ASME. [Online]. https://www.asme.org/topics-resources/content/a-digital-twin-of-a-medical-patient
23. Sun T, He X, Li Z (2023) Digital twin in healthcare: recent updates and challenges. Digit Health 9:20552076221149651. https://doi.org/10.1177/20552076221149651. PMID: 36636729; PMCID: PMC9830576
24. Mayo Clinic Takes the First Step Towards Enterprise-Wide Building Digitizationl. [Online]. http://zsllc-us.com/newsletters/mayo-clinic-takes-first-step-towards-enterprise-wide-building-digitization/

25. Sahilbhutada Case Studies and real-world examples of digital twin implementation. Medium, 28 May 2023. [Online]. https://medium.com/@sahilbhutada28/case-studies-and-real-world-examples-of-digital-twin-implementation-fa650eed06c0
26. Collaborative Research: SCH: A Causal AI Digital Twin Framework to Transform Intensive Care Delivery, Mayo Clinic. [Online]. https://mayoclinic.elsevierpure.com/en/projects/national-science-foundation-award-9
27. Smit JM, Krijthe JH, van Bommel J, van Genderen ME, Labrecque JA, Komorowski M, Gommers DAMPJ, Reinders MJT (2023) The future of artificial intelligence in intensive care: moving from predictive to actionable AI. Intensive Care Med 49(9):1114–1116
28. Zhong X, Babaie Sarijaloo F, Prakash A, Park J, Huang C, Barwise A, Herasevich V, Gajic O, Pickering B, Dong Y (2022) A multidisciplinary approach to the development of digital twin models of critical care delivery in intensive care units. Int J Prod Res 60(13):4197–4213
29. Halamka J Can digital twins improve patient care? Mayo Clinic Platform, 29 Sep 2022. [Online]. https://www.mayoclinicplatform.org/2022/07/12/can-digital-twins-improve-patient-care/

Chapter 8
Future Directions of Digital Twin Technology

8.1 Introduction

The idea of a "digital twin"was developed about 15 years ago. That is a considerable amount of time when compared to the recent explosion of digital technology, which began barely two decades ago. Despite this, it has remained unexplored and has just recently drawn interest from computer science researchers because to its manufacturing roots. Computer Aided Design (CAD) is the source of digital twins; specifically, CAD is used in the mechanical industry to design parts or components. For example, the manufacturing industry uses computer aided design (CAD) software even while creating mechanical parts like cogs (A rotating, circular machine part with cut teeth).

The new generation of information technology has greatly advanced due to the quickly developing fields of computer science and information technology. Some common technologies include artificial intelligence, big data, cloud computing, high-performance computers, and the Internet of Things. Edge computing and cloud computing are used in tandem. Intelligent manufacturing is replacing traditional manufacturing in the sector. Traditional manual operation of equipment is replaced with intelligent maintenance. A new generation of intelligent, digital equipment production and maintenance technology is becoming derivable from it. How to approach flawless interaction data and a high level of integration between real world physical entities and virtual digital items is the fundamental building block. Within this framework, academics have increasingly focused on digital twins as a powerful tool and technical method of facilitating the integration and interaction between real-world hardware and virtual environments. Digital twin technology has already shown tangible progress across numerous industries, demonstrating its significance and transformative potential. Digital twins in manufacturing enhance production scheduling, forecast equipment failures, and increase output. It guarantees virtual patient modelling and tailored medicine in the healthcare industry, which enhance diagnosis and treatment outcomes while lowering associated costs. Digital Twins

© The Author(s), under exclusive license to Springer Nature Switzerland AG 2025

S. Gupta et al., *Digital Twins*, https://doi.org/10.1007/978-3-031-76564-3_8

maximize city services and infrastructures for sustainability and efficiency in urban planning. This ensures, in aerospace and defence, predictive maintenance, and in agriculture, secures precision farming by maximizing resource utilization and increasing production. Considering that it is one of the drivers of sustainability, efficiency, and innovation among various industries, and digital twin technology has many transformational effects today, there is huge potential for even further changes in upcoming years [1].

8.2 Emerging Trends in Digital Twin Technology

The most exciting emerging trend with the digital twin technology has to do with such seamless integrations of AI and machine learning algorithms, further heightening predictive capabilities to hitherto unimagined extents. AI-empowered digital twins lay out predictions for behaviour, detect anomalies, and optimize performance within dynamic environments through vast datasets and real time inputs. This way, it would shift the paradigm in predictive maintenance in manufacturing and further stretch the applications of digital twins into very diversified industries such as healthcare, construction, and urban planning. More importantly, the move into these new industries is a paradigm-changing one to unleash a wealth of possibilities in areas ranging from improving patient care and enhancing building design to developing better urban infrastructure. Also, sophisticated techniques for simulation and modelling, which involve increasingly realistic depictions of physical systems in the virtual world, are being developed side by side with these innovations. In this regard, it can be said that with the ever-increasing growth in digital twin technology being already at a phase of expansion, and with its current sophistication for modelling purposes, its future integration will be combined with AI to transform activity in all spheres with innovation and efficiency like never before [2].

8.2.1 Integrate AI and Machine Learning into Digital Twins for Enhanced Ability in Prediction

The use of artificial intelligence and machine learning together in a digital twin has been a revolution in enhancing the technology of digital twin by empowering it with prediction capabilities to an unimagined scale. Through the process of constant learning from new data, AI-driven digital twins realize better pattern recognition, enabling accurate predictions and thus completely change predictive maintenance in manufacturing, among other applications. There is a mix in there that is operationally efficient, cost-effective, yet extends digital twins to domains such as health, construction, and urban planning. In an unchanged manner, this very trend decisive turns toward proactive decision-making and prescriptive problem-solving driving innovation across industries. Further development of AI and ML algorithms and their integration in the framework of a DT will surely open new opportunities to be ahead of challenges and grasp the opportunities in an increasing complex and connected environment for management. New applications can be found in industries beyond manufacturing, which include sectors such as healthcare, construction, and urban planning.

One of those moments in history, better understanding and interplay with complex systems, was the transfer of digital twin technology beyond manufacturing into domains like healthcare, construction, and urban planning. Those could be sectors benefited from digital twins in the building of virtual replicas. This would provide a complex, real-world environment that is completely mirrored for analysis, simulation, and optimization. Digital twin patients and medical facilities support individual treatment plans, predict diagnostics, and enable optimizations in hospitals. The digital twin buildings and infrastructure in construction make designs easier, promote better project management, and grant sustainability through lifecycle assessments. Furthermore, it enables urban planning to model traffic flow far more efficiently and to optimize resource supply chains for the future with the help of a digital twin city, to plan growth more precisely. It is a paradigm shift toward data-driven decision-making, encouraging innovation, resilience, and sustainability in different industries. The probability of digital twin technology revolutionizing the capability of transforming these sectors into a smarter and efficient system will increase if it continues to develop further and gets linked with other innovative technologies such as Artificial Intelligence and IoT.

8.3 Advanced Simulation and Modelling Techniques Development

Advanced techniques in simulation and modelling are a critical frontier for the development of digital twins, by which more accurate and dynamic depictions of real-life systems would be developed. Advanced simulation and modelling techniques enhance the predictive accuracy and cross-domain versatility of DTs by elaborating things such as material properties, environmental conditions, complex interactions, etc. Some of the advanced techniques in manufacturing provide elaborate process simulations to support process optimization, troubleshooting, and mitigation of risks. In the case of healthcare, they enable patient-specific models for diagnostics, treatment planning, and drug development. Advanced simulation and modelling methods underpin the assessment of a spectrum of scenarios regarding urban planning, infrastructure management, and any decision-making as it touches on questions of sustainable development, disaster resilience, and fitting resource allocation. With these techniques, due to the advancement of computational power, data analytics, and machine learning, digital twin technology holds great promise in informing smarter, more efficient, and resilient systems across different industries, placing virtual simulations at the forefront of decision-making and innovation [3].

8.3.1 Types of Digital Twins

Digital twins can model intricate individual components as well as large-scale systems because to their built-in sophisticated simulation capabilities. While "to replicate an existing physical entity into a digital format" is the common goal of all digital twins, their functions are distinct and they serve diverse purposes. Now let's examine the four most prevalent (as shown in Fig. 8.1) kinds of digital twins.

1. Component Twins: Digital twins that simulate individual portions of systems or products, such as mechanical or electronic components, enabling engineers and developers to digitally design and study the elements. With the addition of maintenance and overall product dependability forecasts, this will assist businesses in simulating their behaviour, performance, and interactions to reach optimal design.
2. Asset twins: These appear as digital twin representations of complete real assets or objects, such buildings, machinery, or even full facilities. With data collected from the asset's development to the decommissioning stage, they provide an all-encompassing picture of its entire cycle. Thus, an asset twin assists a business in remotely monitoring and controlling an asset, tracking performance, and making data-driven decisions on upkeep, effectiveness, and lifecycle management. They

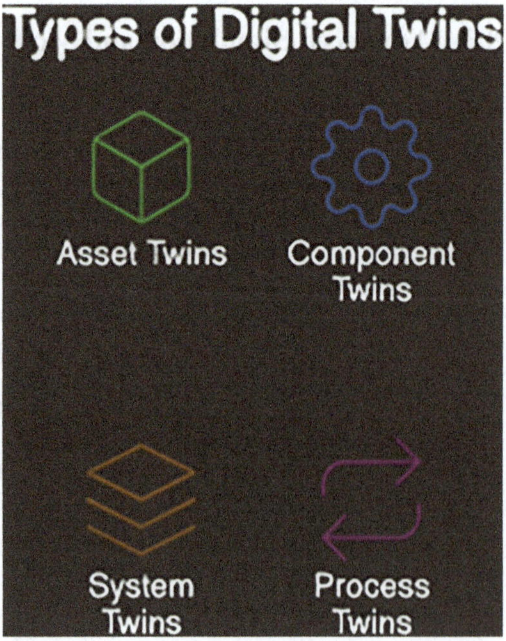

Fig. 8.1 Types of digital twin

are especially helpful for sophisticated assets like infrastructure, buildings, and industrial machines.

3. System twins: Among the systems are digital twins, which are exact duplicates of networked systems such as smart cities, transportation networks, and electricity grids. To provide an all-encompassing picture of these systems' functionality, they combine data from several sources, such as sensors and Internet of Things devices. System twins offer predictive maintenance, real-time monitoring, and performance improvement. By doing this, businesses may enhance the resilience of essential infrastructure, prevent operational instability, and comprehend system-level performance.
4. Process Twins: Digital replicas of an organization's whole workflow or set of processes are known as process twins. They imitate the information, material, and activity flow of a specific process in question to help firms identify bottlenecks, determine the most effective way to allocate resources, and increase process efficiency overall. Streamlining such complicated processes is certain to yield cost savings and operational benefits, which is why businesses like logistics, supply chain management, or healthcare would benefit from the use of process twins.

8.4 Impact of Digital Twin Technology on Society and the Environment

The digital twin technology has huge potentials to transform the society and set it free from many environmental challenges considering the efficiency, productivity, resource use and ethical considerations it inflicts.

It is capable of potentially surging efficiency, productivity, and resource use in many industries, hence positive impacts exerted on society and the environment. Digital twinning is a virtual replica of devices and processes that performs real-time monitoring, analysis, and optimization to save much-needed resources by smoothing the operations. The principal objectives of the digital twin technology in manufacturing are the optimization of production workflows, reduction of downtime through predictive maintenance, and improvement in energy efficiency by detection of highly focused areas. It also carries out specified farming techniques in agriculture with optimized water and fertilizer usage to receive maximum crop yields. Within transportation and logistics, digital twins focus on route optimization, reduction of fuel expenditure, and lowering emissions to achieve environmental sustainability. Savings in costs engendered by such efficiency gains will go hand in glove with more sustainable ways of using resources, lessening their impact on the environment and promoting good stewardship over natural resources. Besides, DTs enable individuals and organizations to make informed decisions based on data-driven insights, entrenching a culture of constant improvement and innovation that possibly will contribute to heterogeneous positive societal and environmental results. All in all, the adoption of digital twin technology is going to totally revolutionize the way we engage with our environment, leading to higher efficiency, sustainability, and resilience of systems for the betterment of society and the planet [4].

Positive Effects on Efficiency, Productivity, and Resource Usage Digital twins allow for real-time process monitoring and, hence, are optimally efficient across different industries, improving efficiency and productivity. They allow organizations to simulate a range of scenarios concerning physical assets and systems and, therefore, enable the detection of inefficiencies in order to enhance them. This reduces lag time while optimizing resource usage and smoothening operations. Moreover, the creation of digital twins realizes predictive maintenance, thus making it possible to make proactive interventions for the prevention of equipment failure and minimize disruptions. This will enable an organization to maximize the life of its assets while minimizing the maintenance cost and enhancing resource utilization for sustainability.

8.4.1 Potential of Reducing Environmental Footprint Through Optimized Processes and Predictive Maintenance

Another high-tech contribution that digital twin technology can give right off the peg is assistance towards the lessening of environmental impact through optimization of processes and predictive maintenance strategies. As it can read and analyse real-time information, organizations can make out opportunities for minimizing energy use, generation of wastes, and emissions. Digital twins can optimize business manufacturing processes, supply chain logistics, and energy management systems to achieve reduced consumption of resources and reduced environmental pollution. Moreover, predictive maintenance capabilities prevent sudden breakdowns of equipment, which may lead to huge loss of energy and cause environmental pollution. Digital twins save resources and are environmentally friendly since they keep the equipment in good condition and avoid any extra downtime.

Among others are:

1. Optimized Operations: Digital twin technology makes it possible to monitor and analyze, in real-time, processes for inefficiencies in business operations. Therefore, it will increase productivity and reduce wastage, hence efficient usage of resources.
2. Predictive Maintenance: Digital twins allow for the creation of virtual replicas of physical assets, hence developing predictive maintenance that prevents equipment failure and reduces downtime. In this way, it enhances operational efficiency by elongating the life span of these assets, consequently reducing the frequency of replacements to a minimum and minimizing the use of resources.
3. Resource Conservation: By using digital twins, resource usage, like energy, water, and raw materials, can be tracked and controlled to a large extent. Their

usage, after optimization, reduces wastes and the after-effects on the environment, thereby conserving resources.
4. Improved Product Design: Digital twins can be used to test and prototype products virtually, therefore allowing companies to fine-tune designs before deciding on specifications for physical production. Since this is an iterative process, it reduces the loss of material and energy consumption that normally results from trial-and-error methods applied in the manufacturing process.
5. Optimizing Supply Chains: Digital twins provide insight into the dynamics of the supply chain, which enables companies to optimize routes for transportation, inventory levels, and distribution networks. By optimizing processes within the supply chain, energy use and emissions from logistics operations are lowered.
6. Urban planning and infrastructure: Digital twins of cities and infrastructural assets enable better urban planning decisions. Running simulations and analyzing data will allow city planners to design more sustainable transport systems, energy grids, buildings, and all those other flaws that exist in a city to minimize its carbon emissions and enhance quality of life for its citizens.
7. Environmental Monitoring and Management: Digital twins can be tracking environmental parameters—such as air and water quality, biodiversity, and the health of ecosystems. This would be possible through the integration of sensor data and predictive modeling, enabling environmental agencies to make informed decisions for the protection and restitution of natural resources.
8. Ethical Considerations on Data Ownership, Transparency, and Equitable Access:

The application of digital twin technology also brings to the fore some important ethical concerns with respect to data ownership, transparency, and access. In terms of the huge amount of data to be collected and analyzed for digital twins, there are concerns about data privacy provisions and security measures, not mentioning the ownership rights over this data. It will be important to understand how organizations can have transparent data governance frameworks and mechanisms in place to seek informed consent from the use of data by people whose data is being collected and utilized. A good deal of attention should be given to the reduction of the digital divide in accessing digital twin technology and the ways through which its benefits can be distributed equitably among different socio-economic groups or geographic regions. The above ethical considerations will, therefore, help stakeholders build trust, accountability, and inclusivity into the development and deployment of solutions for digital twins [5].

8.5 Future Predictions and Trends for Digital Twins

In the future, the following technological issues must be resolved to create a developed digital twin model of complicated machinery and encourage the use and use of digital twin technology.

1. Creating a high-fidelity, multi-scale, multi-physics field linked model: Large-scale machinery frequently consists of intricate mechanical, electrical, hydraulic, gas, and thermal components. The description of various physical field scales varies, making the establishment of high-precision simulation models exceedingly challenging.
2. Interaction between physical things and twin models in real time: To achieve the real-time updating of twin models, many sensors must be deployed in order to record temperature, speed, flow rate, and pressure. It also involves using computer twin models to manipulate physical elements.
3. Large-scale data processing and high-performance computing: As communication and sensor technologies have advanced, so has the volume of data being collected. This raises the bar for computers' capacity to handle data and do computation.

The idea of digital twins has changed over the past 10 years. Stage 1 involved modelling a physical entity, which was mostly used during the design phase. Stage 2 involved mirroring an existing physical entity, which could be used as a reference model for a particular physical entity. The concept of "instance" is introduced here; at stage 1, nothing existed; a digital twin would represent a model of a turbine, not a particular turbine that was produced. The Thread component, which documents the development of that specific instance—how the physical entity was created, developed, and sold—has been included to the Digital Twin.

Stage 2.5 of that evolution resulted in replicating the physical object's current state. Since different instances of a product will be in different states depending on how they are being utilized, the concept of instance is crucial here. A physical entity's associated instance can be synchronized with it using a variety of methods, from straightforward data entry to a constant stream of data from the physical entities themselves. Typically, this is done via sensors and IoT embedded in the physical entity and connected to the digital twin via a network. With the addition of a Digital Shadow component, the three elements of modern digital twins—a digital model, a digital shadow, and a digital thread—were completed.

There is now a two-way exchange between the physical thing and its Digital Twin during stage 3, when the majority of the industry is using Digital Twins. This led to the creation of an enlarged Digital Thread, which now keeps track of all significant operational events as well as the potential to influence the physical entity's functioning autonomously. With this, the idea of a digital twin serving as a passive stand-in for a physical object is eliminated.

Nevertheless, once it is in place, the digital twin can impact the original physical object and even have the ability to function as an independent active element. These days, the digital twin "influences" the physical twin mostly by serving as a conduit through which an external controller can interact with the physical twin. The Digital Twin made no independent decision that led to this contact. The digital twin is then advanced to stage 4, when there is some autonomy between the two sets of digital and physical twins.

At this point, the interworking of the digital and physical twins offers some entity functionality; taken as a whole, they establish local data analytics with a minimal degree of intelligence at the digital twin level. At stage 5, Digital Twins gain more autonomy to the point where they can engage in cyberspace as independent agents, bringing local analytics to a global scale. The stages is as shown in Fig. 8.2. An emergent ambient intelligence not shown in the graphic will develop from the coming together of self-generation of digital twins, hierarchies of digital twins, and mesh networks of digital twins [6].

If we look at more general application areas, the evolution has been exponential: starting with early manufacturing applications that made use of digital models created during the design phase, they have been adopted in construction, automotive, aerospace, and cities; more recently, they have been used in agriculture, healthcare, and energy. The expansion is still running strong, and it is currently making a significant shift to the modeling of people, information, economy, finance, immaterial goods, and processes.

The latter is an inevitable byproduct of resource management and healthcare that is beginning to affect education, especially new paradigms for continuous learning. Thanks to the rise in the quantity and caliber of sensors as well as the corresponding rise in the volume and quality of data, there has been a horizontal growth with a

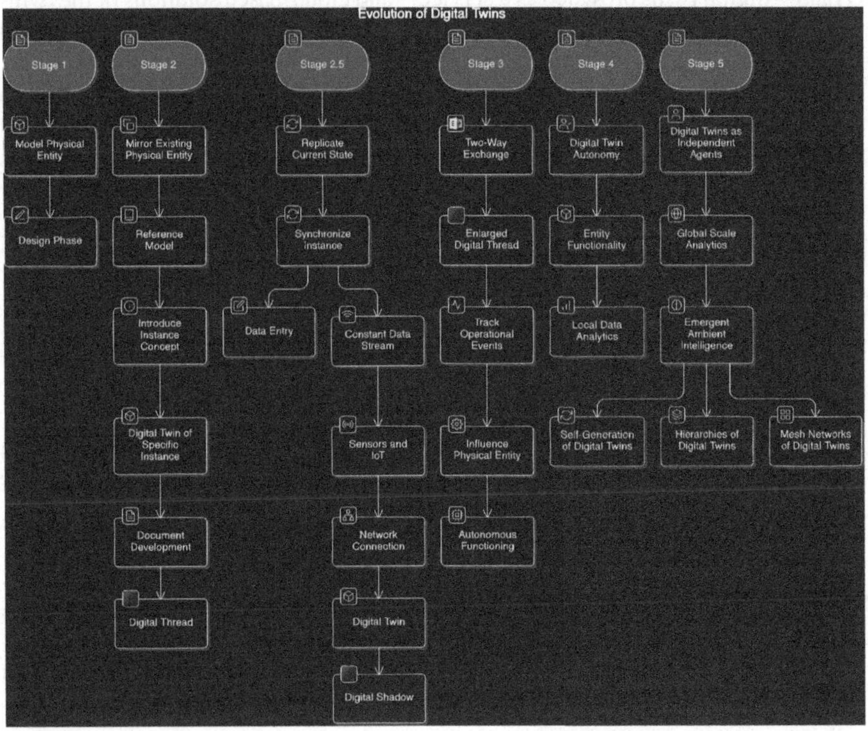

Fig. 8.2 Evolution of digital twin

greater range of applications, while the vertical expansion is associated with the coverage of physical entities.

8.6 Digital Twin Entry into New Industries and Sectors

Digital twins are bound to expand their traditional domains and penetrate new industries and sectors. Whereas the digital twins have earlier been adopted into sectors like manufacturing and aerospace, with time they find use in healthcare, transport, and retail, among many others. They will be used to monitor the health status of a patient, optimize supply chains, manage energy grids, and improve customer experiences, among many others. Being versatile and adaptable, digital twins will become a perfect tool in finding a solution to intricate problems and drive innovation across various dimensions of any industry.

8.6.1 Digital Twins as a Service and Cloud-Based Solutions

DTaaS—digital twins provided as a service—and cloud-based solutions of the twins will be placed within the digital twin continuum. It democratizes access to digital twin technology, making it much easier and more affordable for any organization, regardless of size. DTaaS models will mean scalable and flexible solutions that no longer require huge investments in infrastructure. Cloud-based platforms will ensure that data storage and processing capability are safe, with the ability to access them from any part of the world. It allows organizations to embrace the full power of the digital twin without worrying about complexities in IT system management.

8.6.2 Digital Twins Are Coming to Converge with Extended Reality (XR)

Experiences created by the integration of Digital Twins with Virtual and Augmented Reality Extended Reality technologies make it possible to interact and user immerse themselves within.

Digital twin XR interfaces will allow intuitive visualization and interactivity, increasing understanding, collaboration, and decision-making. For example, engineers, operators, and maintenance technicians might use a digital twin to overlay information on top of the physical asset that would guide and assist in real-time. Where XR visualization can seamlessly bridge the gap to integrate the physical and digital worlds, which in turn revolutionizes training, troubleshooting, and remote collaboration.

8.6.3 Integration of Edge Computing for Real-Time Analytics

In the future of digital twins, edge computing capabilities will be integrated to ensure real-time analytics at the edge of the network. Based on this assumption, it processes data closer to the source; hence, fast response to the changing conditions and gives immediate insights.

Edge computing is going to reduce latency, improve data security, and allow efficient utilization of network bandwidth. This development shall be underlined in scenarios having real-time decision-making including autonomous cars, smart cities, and industrial automation.

8.6.4 Digital Twin Technology Data Privacy and Moral Considerations

It will come to the fore when it is pervasive in nature. Organizations and policymakers will have to satisfy arguments related to ownership of information, its secrecy, and consent in this regard. For this, clear guidelines and regulations on the responsible and ethical use of DT technology will be required. This shall be further supported by important paradigms like anonymization of data, encryption, and proper control of access to technical limitations in ensuring the protection of sensitive details and people's privacy.

Balancing the use of data with privacy protection will become key to winning public trust and continuing that base for which DT technology is to achieve sustainable adoption. Integration of 5G connectivity to support seamless data transfer Massively deployed 5G networks are about to change the full potential of digital twin technology. Having incredibly fast, low latency, and enormous connectivity, 5G is supposed to sustain the smooth data exchange between digital twins and their physical prototypes. This will involve real-time synchronization, remote control, and monitoring for assets and processes. Due to the connectivity developed and responsiveness of 5G, new opportunities related to innovation and efficiency shall lie in autonomous transport and smart manufacturing, and healthcare. Sustainability and Environmental impact The Digital twin technology shall increasingly take up the challenge of sustainability and respond to the environmental challenges also. Enterprises will make use of digital twins to reduce energy consumption, structure carbon emissions, and heighten resource efficiency.

Digital twins have been used in designing eco-friendly products, sustainable city planning, and predictive maintenance. The potential for simulation and analysis that the digital twin can turn out for the achievement of sustainability goals towards a greener future lies in lowering the environmental footprint of systems and processes. Key Insights and Way Forward for Digital Twin Technology Digital twin technology has been organically evolving into a power agent with enormous ramifications across industries and sectors of society. The continuous development and traction gained from its current application would reflect a variety of trends and opportunities that would shape its main trajectory in the future. Advanced Technologies Integration: The future will be characterized by further linkage between digital twin technology and advanced technologies such as Artificial Intelligence, Machine Learning, and the Internet of Things. Resultantly, this will amplify the predictive abilities where twins will present skewed simulations, proactive decision-making, and autonomous operation across a broad domain [7].

Entry into New Industries: Because digital twins were primarily used in industries such as manufacturing and engineering, with increasing awareness, entry is expected to grow in completely new fields like healthcare, construction, agriculture, and smart cities. Such industries will profit from the insights and thus the optimization afforded by digital twins as a means of pushing innovation and efficiency into traditionally analog processes. Next-generation digital twin solutions will be equipped with advanced real-time monitoring and control that enables an organization at chin speed to respond to the changing conditions and optimize performance in dynamic conditions. This would be very useful in agile decision-making and for enhanced operational resilience, hence improving overall system reliability. Interconnected Ecosystems of Digital Twins: Having found a lot of applications, the most adopted development we should witness in this decade for digital twin technology will be an interconnected ecosystem of digital twins—where several digital replicas shall be able to communicate and exchange data in simulating complex systems and ecosystems. That would be holistic modelling and optimization of interconnected systems like smart cities or interlinked supply chains. Focus on Sustainability and Resilience: With growing awareness of ecological and social issues, digital twin technology will increasingly be at the heart of driving sustainability and resilience. This is where organizations will work toward the optimal use of resources, low ecological footprint, and improved resilience of important infrastructures to disruptive events in a changing world, like climate change and pandemics. Ethical and Regulatory Considerations: As adoption increases, so will digital twin technology zero in on their ethical and regulatory considerations around data privacy, security, and governance. In this respect, stakeholders are called upon to provide robust frameworks that entrench the governance of data through either transparency and accountability in their usage or principles of fairness and equity in access dissemination to digital twin solutions.

8.7 Conclusion

Prospects for digital twin technology in the future remain very bright, as they may transform whole industries and improve societal welfare while fighting against environmental threatsWhat clearly emerges from already realized or on-going development is that, within a wide array of domains, Digital Twins do give unprecedented competencies related to process optimization, outcome prediction, and finally, Innovation. Be it in the domains of manufacturing and health or smart cities and agriculture, the digital twins are all set and poised to bring in a sea of change in current traditional practices and usher in a new era for efficiency, productivity, and sustainability. Most of the major insights that do come from observation into the future are those where the integration of digital twins with cutting-edge technologies would further their use and probable effect in a good number of differing sectors. Digital twin interoperability in ecosystems allows holistic modeling and optimization of complex systems for agility, resilience, and innovation. The second is that stronger consideration of sustainability and ethical dimensions will drive the responsible development of solutions for digital twins to ensure equitability in benefit distribution and alignment with the values of society. In that direction, more than the joint effort put into resolving such issues as data privacy, interoperability, and compliance, it would be very collaborative among stakeholders, very transparent and accountable, and strong in governance arrangements so that these challenges can be surmounted and unlock, once and for all, the potential of digital twins for societal and environmental good. Bright prospects await digital twin technology, hopping with endless scope for innovation and progress. Embracing the potentials of digital twins and related challenges shall unleash new horizons of knowledge, efficiency, and sustainability and thus create a better future for humanity.

References

1. https://insights.daffodilsw.com/blog/the-future-of-digital-twins
2. Zhang Y, Fang L, Deng H, Qi Z, Liang H (2023) Recent advances and future perspectives of digital twins. In: IEEE international conference on control, electronics and computer technology (ICCECT), Jilin, China, 2023, p 1563–1566. https://doi.org/10.1109/ICCECT57938.2023.10140652
3. https://cmte.ieee.org/futuredirections/2021/01/26/digital-twins-future/
4. Alnowaiser KK, Ahmed MA (2023) Digital twin: current research trends and future directions. Arab J Sci Eng 48:1075–1095. https://doi.org/10.1007/s13369-022-07459-0
5. Tao F, Zhang H, Liu A, Nee AYC (2019) Digital twin in industry: state-of-the-art. IEEE Trans Industr Inform 15(4):2405–2415. https://doi.org/10.1109/TII.2018.2873186
6. Crespi N, Drobot AT, Minerva R (2023) The digital twin in action and directions for the future. In: Crespi N, Drobot AT, Minerva R (eds) The digital twin. Springer, Cham. https://doi.org/10.1007/978-3-031-21343-4_38
7. Liu J, Duan L, Lin S et al (2024) Concept, creation, services and future directions of digital twins in the construction industry: a systematic literature review. Arch Computat Methods Eng. https://doi.org/10.1007/s11831-024-10140-4

Index

© The Editor(s) (if applicable) and The Author(s), under exclusive license to Springer Nature Switzerland AG 2025
S. Gupta et al., *Digital Twins*, https://doi.org/10.1007/978-3-031-76564-3

SPRINGER NATURE

GPSR Compliance

The European Union's (EU) General Product Safety Regulation (GPSR) is a set of rules that requires consumer products to be safe and our obligations to ensure this.

If you have any concerns about our products, you can contact us on ProductSafety@springernature.com

In case Publisher is established outside the EU, the EU authorized representative is:

Springer Nature Customer Service Center GmbH
Europaplatz 3
69115 Heidelberg, Germany

The manufacturer's authorised representative in the EU is Springer Nature Customer Service Centre GmbH, Europaplatz 3, 69115 Heidelberg, Germany. If you have any concerns regarding our products, please contact ProductSafety@springernature.com

Printed and bound by CPI Group (UK) Ltd, Croydon, CR0 4YY
28/09/2026
02224081-0005